Keys to the Vertebrates
of the
Eastern United States
Excluding Birds

by

JOHN O. WHITAKER, Jr., Ph.D.

Indiana State University
Terre Haute, Indiana

Burgess Publishing Company

426 South Sixth Street • Minneapolis, Minn. 55415

PREFACE

These keys were originally prepared for use in Vertebrate Systematics and Ecology Courses at Indiana State University, Terre Haute. All sections of the keys have been used for at least one year, while other sections have received up to five years of use under class conditions.

The keys were written because other available keys were local in nature or restricted taxonomically, or else treated such a large geographic area and such a large number of species that undue difficulty was encountered during their use.

It is hoped that the present keys will enable college students in courses in vertebrate systematics or ecology, natural history, or field biology or zoology to identify the vertebrates presently existing in the eastern United States. An effort has been made to include all species, yet to keep the keys as simple as possible and still allow the student to identify the majority of adult specimens.

Distributional information and keys have been obtained from many sources, too numerous to mention. Some of the keys are in their original form and some have been greatly modified. Much credit must be extended to Blair, *et. al.* (1957. Vertebrates of the United States) for without this work the keys to the fishes probably would not have been possible. Many of the range maps were drawn from previously published maps, while others were drawn from word descriptions.

Several individuals have greatly aided during the preparation of these keys. I here express my sincere appreciation to all of them. Dr. J. Knox Jones, Jr. and some of the Graduate students at the University of Kansas have read and criticized various sections. David Rubin has read many parts of the manuscript, and has helped in searching the literature and in final organization of the keys. My wife, Royce, spent many hours helping in the preparation of the range maps. Barbara Fox, Judi Boyd, Georgia South, Reba Griffin and Sharon Rainbolt have helped in typing various drafts of the manuscript.

Finally, the members of my classes have utilized, criticized and tested the keys. It is because of them that many errors and "difficult couplets" have been eliminated or improved.

INTRODUCTION

These keys have been prepared as an aid in the identification of vertebrate animals, excluding birds, in the United States east of the Mississippi River.

An attempt has been made to include all species currently recognized in that area, and to utilize external, easy-to-use characteristics. A range map has been included for each species so that the student can see whether the species to which he has keyed his specimen is one that might reasonably be expected to occur in the area under consideration. If an unreasonable answer has been obtained in terms of geographic range he should probably try keying the specimen again. If he still arrives at the same answer, or if his answer is a reasonable one in terms of geographic range, then he should read the species description in a more comprehensive work such as one of the selected references listed on page 245 or the sources in the general bibliography at the end of the book.

It would be ideal if simple keys could be written that would allow one to identify organisms unfailingly. Unfortunately, there is great variation among animals even within species, making it very difficult or impossible to construct keys that are simple, but that still work for all individuals. Keys intended to work for all individuals would be so lengthy and hard to use that they would lose much of their value. It is hoped, however, that the great majority of specimens can be identified using these keys. The use of series of specimens rather than of individuals may sometimes mean greater success in the use of the keys.

At the beginning of each vertebrate section are instructions and diagrams concerning the major characters to be used. Most unfamiliar terms will be defined and explained here. The student should study these sections carefully because a thorough understanding of this material will save him time later on. The most benefit could probably be obtained by examining specimens at the same time that the definitions and figures are studied.

TABLE OF CONTENTS

STRUCTURES OF FISHES

Study this section and figures 1-3 carefully.

Adipose fin. Fleshy fin without spines or rays behind the dorsal fin (Fig. 1).

Barbel. Fleshy protuberance, usually near the mouth, and usually quite small (Fig. 1).

Body depth. Measure at deepest point.

Branchiostegal rays. The elongate bones supporting the membranes of the posterior ventral portion of the head just below the gill covers.

Caudal peduncle. From posterior base of anal fin to end of vertebral column (Fig. 1).

Ctenoid scale. A scale with tiny spines on its posterior portion.

Cycloid scale. A scale, roughly round or oval, and lacking the spines of ctenoid scales.

Falcate. Usually refers to dorsal fin; curved or sickle shaped. The posterior edge is concave, rather than straight or convex.

Fimbriate. Edge with fringe of elongate processes.

Frenum. Fleshy bridge connecting premaxilla to snout, preventing the free movement of premaxilla from snout.

Ganoid scale. Very hard scales which fit closely together like tiles on a floor. They are arranged in diagonal rows.

Gill rakers. The anterior projections from the gill arch. Can be seen by pushing the operculum forward.

Head length. Tip of snout or jaws to posterior edge of gill cover (Fig. 1).

Heterocercal tail. The posterior vertebrae bend upward, thus entering the upper lobe of the caudal fin.

Homocercal tail. The vertebrae end at the base of the tail.

Hypural plate. End of the vertebral column. Locate by bending tail from side to side.

Gill membranes. These lie in the area of the throat of the fish, may or may not be connected across the isthmus.

Isthmus. The ventral portion of the fish's throat lying between the gill openings.

Lateral line. Series of tubes and pores on the side of a fish. May be complete, running from gill cover to caudal fin or may be incomplete. To count scales in lateral line count from the first scale over the shoulder girdle to the scale over the hypural plate.

Myomeres. Muscular segments that can be seen laterally between gill openings and anus.

Opercular membranes. Membranes below and behind the operculum.

Palatine teeth. (Fig. 3).

Papillae. Small fleshy projections.

Papillose. With papillae.

Peritoneum. Lining of the body cavity.

Pharyngeal teeth. When working with the minnows it is imperative that one knows how to dissect out the pharyngeal teeth. These are sometimes the most certain means of identification of species. With practice it is possible to take out the pharyngeal arch of a fish, yet have a "good" specimen for preservation. To remove the pharyngeal teeth hold the fish in one hand, and with the other, push the operculum forward, exposing the gill arches. Then push forward the gill arches with a thumb or dissecting needle. Cut through the tissues behind the gills, allowing them to be pushed still farther forward. With forceps then grasp the pharyngeal arch and gently pull it out, being careful not to injure the arch or the teeth on it. Next clean the tissue from the teeth using dissecting needles.

Plicate. With longitudinal folds.

Postorbital length of head. Back of eye to end of gill cover.

Predorsal length. Tip of snout or jaw to anterior base of dorsal fin.

Premaxillary teeth. Most anterior group of teeth of upper jaw.

Premaxilla protractile. Unobstructed groove present behind the premaxilla; premaxilla can be pulled out from the snout with forceps or needle.

Preopercle. (Fig. 2).

Preoperculomandibular canal. The canal from the anterior end of the lateral line which crosses the preopercle and extends onto the mandible.

Pseudobranchia. Small patch of gill filament material on the inner surface of the operculum.

Pyloric caecae. Branches from gut at base of intestine.

Rays. With crosswalls, usually branched and usually soft (Fig. 1).

Retrorse. Curved backward.

Snout length. Snout to eye.

Spines. No crosswalls, no branching, usually stiff (Fig. 1).

Spiracle. An opening; in elasmobranchs the opening between the eye and the gill opening.

Subopercle (Fig. 2).

Subterminal. Some distance from the end.

Supraopercle. (Fig. 2).

Villiform bands (teeth). Close together in bands, of equal lengths, and slender finger like in shape.

Vomer (Fig. 3).

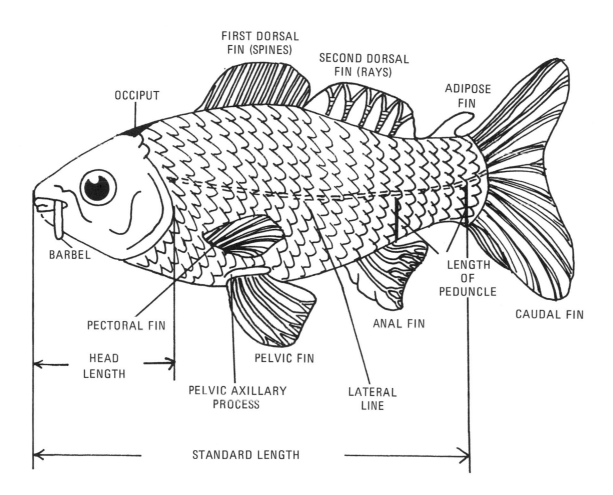

FIRST DORSAL FIN (SPINES)

SECOND DORSAL FIN (RAYS)

ADIPOSE FIN

OCCIPUT

BARBEL

PECTORAL FIN

PELVIC FIN

PELVIC AXILLARY PROCESS

HEAD LENGTH

LATERAL LINE

ANAL FIN

LENGTH OF PEDUNCLE

CAUDAL FIN

STANDARD LENGTH

END OF VERTEBRAL COLUMN (FIND BY BENDING TAIL FROM SIDE TO SIDE.)

Figure 1
EXTERNAL STRUCTURE OF A FISH

4

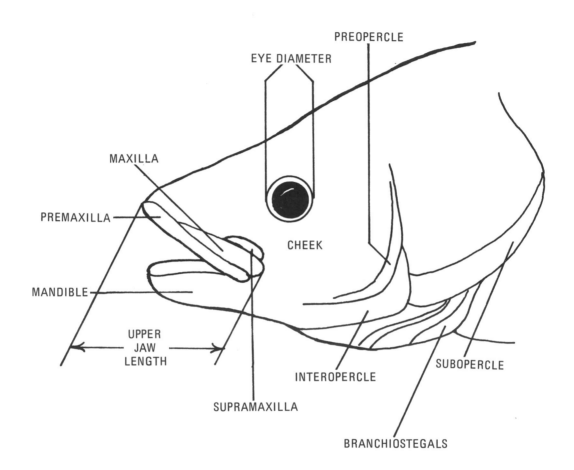

PREOPERCLE

EYE DIAMETER

MAXILLA

PREMAXILLA

CHEEK

MANDIBLE

UPPER
JAW
LENGTH

INTEROPERCLE

SUBOPERCLE

SUPRAMAXILLA

BRANCHIOSTEGALS

Figure 2
STRUCTURES OF THE HEAD

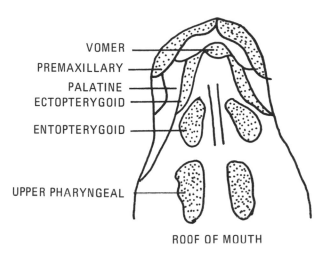

VOMER

PREMAXILLARY

PALATINE

ECTOPTERYGOID

ENTOPTERYGOID

UPPER PHARYNGEAL

ROOF OF MOUTH

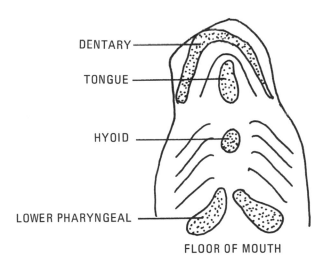

DENTARY

TONGUE

HYOID

LOWER PHARYNGEAL

FLOOR OF MOUTH

NOTE THE LOWER PHARYNGEAL TEETH. THESE ARE THE ONES YOU
SHOULD LEARN TO DISSECT OUT. (SEE UNDER PHARYNGEAL TEETH)

Figure 3
ROOF AND FLOOR OF MOUTH OF FISH
SHOWING TYPES OF TEETH

KEYS TO FRESHWATER FISHES

(Not included are some primarily salt water forms which may be found relatively far up coastal streams)

A. Body eel-shaped, with or without jaws

  B. No jaws, mouth a circular disc . . . . .PETROMYZONTIDAE

    C. Dorsal fin not completely divided . . . . *ICHTHYOMYZON*

      D. Oral disc contained 7.7 - 16.9 times in total length. Teeth well developed. Parasitic forms

        E. Most circumoral teeth unicuspid. Myomeres 49-52 . . . . *I. unicuspis*

        EE. At least six circumoral teeth partly bicuspid. Myomeres 51-58

          F. Myomeres 51-54, bicuspid circumorals 6-8 . . . . *I. castaneus*

          FF. Myomeres 56-58, bicuspid circumorals usually 8 . . . . *I. bdellium*

      DD. Oral disc 15.4 to 27.7 times in total length. Teeth degenerate at least on posterior field of disc. Nonparasitic

        E. Circumoral teeth unicuspid, anterior teeth 2, visible lateral teeth 3-5. Myomeres 50-52. . . . . *I. fossor*

        EE. Circumoral teeth in part bicuspid, anterior teeth 4, lateral teeth 7-8. Myomeres 52-60

          F. Myomeres less than 55 . . . . *I. gagei*

          FF. Myomeres 55 or more

            G. Height of first dorsal fin greater than distance from anterior margin of eye to nostril . . . . *I. greeleyi*

            GG. Height less than this distance . . . . *I. hubbsi*

    CC. Dorsal essentially divided into two fins (but sometimes slightly connected)

      D. Disc teeth large and numerous. Myomeres about 70. Parasitic . . . . *Petromyzon marinus*

      DD. Disc teeth inconspicuous, few. Myomere number varies . . . . . *LAMPETRA*

I. unicuspis
Hubbs and Trautman
Silver Lamprey

I. castaneus
Girard
Chestnut Lamprey

I. bdellium
(Jordan)
Ohio Lamprey

I. fossor
Reighard and Cummins
Northern Brook Lamprey

I. gagei
Hubbs and Trautman
Southern Brook Lamprey

I. greeleyi
Hubbs and Trautman
Allegheny Brook Lamprey

I. hubbsi
Raney
Tennessee Brook Lamprey

P. marinus
Linnaeus
Sea Lamprey

E. Myomeres less than 62. Posterior portion of oral disc with visible teeth other than marginals . . . . *Lampetra lamottei*

EE. Myomeres more than 62. Posterior oral disc lacking teeth other than tiny marginals . . . . *L. aepyptera*

BB. Mouth normal with movable jaws . . . . ANGUILLIDAE, *Anguilla rostrata*

AA. Body not eel-shaped. Jaws present

B. Caudal fin heterocercal or abbreviate heterocercal

C. Caudal fin forked, mouth subterminal

D. Snout long and paddlelike, with 2 minute barbels on ventral surface. Skin naked . . . . POLYODONTIDAE, *Polyodon spathula*

DD. Snout relatively short, with 4 long barbels ventrally, body covered with 5 rows of bony plates . . . . ACIPENSERIDAE

E. Caudal peduncle incompletely armored. Spiracles (over eyes) and pseudobranchia present . . . . *ACIPENSER*

F. Fewer than 15 gill rakers on lower arm of first gill arch . . *A. oxyrhynchus*

FF. Usually 15 or more

G. Anal rays about 37. Gill rakers 14-19 . . . . *A. fulvescens*

GG. Anal rays about 22. Gill rakers 17-27 . . . . *A. brevirostris*

EE. Caudal peduncle completely armored. Spiracles and pseudobranchia absent. Caudal filament present . . . . SCAPHIRHYNCHUS

F. Belly with small dermal plates. Dorsal rays 30-36. Anal rays 18-23 . . . . *S. platorynchus*

FF. Belly largely naked. Dorsal rays 37-43. Anal rays 24-28 . . . . *S. album*

CC. Caudal fin rounded behind. Mouth terminal

D. Scales ganoid. Dorsal short. Snout an elongate beak. . . . . *LEPISOSTEIDAE, LEPISOSTEUS*

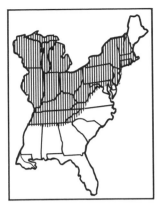

L. lamottei
(LeSueur)
American Brook Lamprey

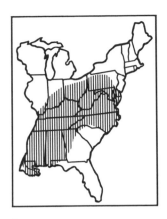

L. aepyptera
(Abbott)
Ohio Brook Lamprey

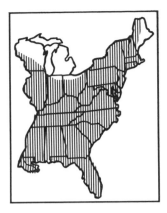

A. rostrata
(LeSueur)
American Eel

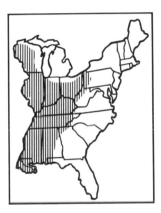

P. spathula
(Walbaum)
Paddlefish

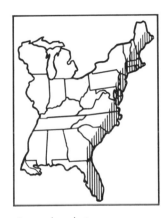

A. oxyrhynchus
Mitchill
Atlantic sturgeon

A. fulvescens
(Rafinesque)
Lake Sturgeon

A. brevirostris
LeSueur
Shortnose Sturgeon

S. platorynchus
(Rafinesque)
Shovelnose Sturgeon

S. album
(Forbes and Richardson)
Pallid Sturgeon

E. Snout very long, width of upper jaw at nostrils less than diameter of eye. Distance from posterior edge of eye to posterior edge of opercular membrane contained more than 3.5 times in head length . . . . . . *L. osseus*

EE. Snout shorter, shaped like duck bill. Width of upper jaw greater than diameter of eye. Distance less than 3.5 times in head length

F. Diameter of eye 1.5 or more times in width of upper jaw at nostrils. Distance from posterior edge of eye to posterior edge of opercular membrane 2.5 to 2.9 in head length. . . . . *L. spatula*

FF. Diameter of eye less than 1.5 times in jaw width . Distance 2.9 to 3.5 in head length

G. Lateral line scales 59-64 (including tiny ones on caudal fin). Predorsal scales 50-54. Scales from anal plate to middorsal line 20-22 . .
. . . . *L. platostomus*

GG. Lateral line scales 54-58. Predorsal scales 46-49. Scale rows 17-19

H. Distance from anterior orbit to edge of opercular membrane less than 2/3 snout length . . . . *L. platyrhincus*

HH. This distance more than 2/3 snout length . . . . *L. oculatus*

DD. Scales cycloid. Dorsal long. Snout blunt . . . . AMIIDAE, *Amia calva*

BB. Caudal fin homocereal

C. Body flattened. Eyes both on same side of body . . . .
. . . . .ACHIRIDAE, *Trinectes maculatus*

CC. Eyes on opposing sides of body

D. Body covered with dermal plates. Snout elongate, tubular. Anal fin rudimentary or absent. . . . . .SYNGNATHIDAE, *Syngnathus scovelli*

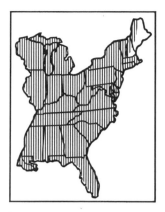

L. osseus
(Linnaeus)
Longnose Gar

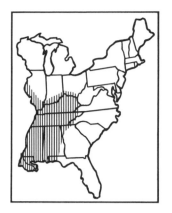

L. spatula
(Lacepède)
Alligator Gar

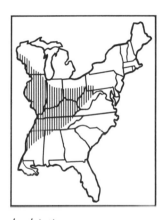

L. platostomus
Rafinesque
Shortnose Gar

L. platyrhincus
DeKay
Florida Spotted Gar

L. oculatus
Winchell
Spotted Gar

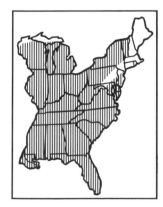

A. calva
Linnaeus
Bowfin

T. maculatus
(Bloch and Schneider)
Hogchoker

S. scovelli
(Evermann and Kendall)
Southern Pipefish

DD. Body scaled or naked, but not with dermal plates. Snout not tubelike. Anal fin present

E. Adipose fin present

F. At least 4, but usually 8 barbels about mouth . No scales . . . . ICTALURIDAE (see key pg. 28)

FF. No barbels . Scales present

G. Scales strongly ctenoid. 1-3 weak splintlike dorsal spines and 1 or 2 anal spines . . . . PERCOPSIDAE, *Percopsis omiscomaycus*

GG. Scales cycloid. No spines

H. Pelvic axillary process present. Usually more than 25 lateral line scales

I. Strong teeth on jaws and tongue . Maxilla extending to below middle of eye . . . . SALMONIDAE (see key pg. 34)

II. Teeth on jaws and tongue weak or absent . Maxilla not extending to below middle of eye . . . . COREGONIDAE (see key pg. 36)

HH. Pelvic axillary process absent. Fewer than 75 scales in lateral line. Mouth large. Maxillary extending beyond middle of jaw . . . . OSMERIDAE, *Osmerus mordax*

EE. Adipose fin absent

F. Prominent single ventral barbel near tip of chin . . . . *GADIDAE, Lota lota*

FF. No single prominent barbel near tip of chin

G. Pectoral fins placed high on the sides, above the axis of the body . . . . BELONIDAE, *Strongylura marina*

GG. Pectoral fins much lower

H. Anal opening in gular region in adults

P. omiscomaycus
(Walbaum)
Troutperch

O. mordax
(Mitchill)
American Smelt

L. lota
(Linnaeus)
Burbot

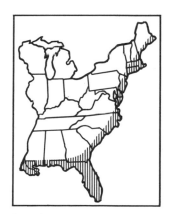

S. marina
(Walbaum)
Atlantic Needlefish

I.   Dorsal spines present. . . APHREDODERIDAE, *Aphredoderus sayanus*

II.   Dorsal spines absent . . . . AMBLYOPSIDAE

   J.   Body pigmented. Eyes well developed, but covered by transparent or translucent skin. Sensory papillae absent . . . . .*CHOLOGASTER*

     K.   9-11 branched caudal rays. Lower sides of body with dark stripes . . . . *C. cornuta*

     KK. 12-16 branched caudal rays. Lower sides without dark stripes
. . . .*C. agassizi*

   JJ. Pigment absent. Eyes absent or degenerate. Sensory papillae in 1 or 2 rows on each half of caudal fin

     K.   Sensory papillae in 1 row on each half of caudal fin . . . . . .
. . . . *Typhlichthys subterraneus*

     KK. Sensory papillae in 2 or 3 rows on each half of caudal fin . . .
. . . . *Amblyopsis spelaea*

HH. Anal opening just before anal fin

   I.   2-15 dorsal spines present, free, not connected by membranes
. . . . .GASTEROSTEIDAE

   J.   Gill membranes joined across isthmus

     K.   Pelvic bones not joined, each extending backward as strong process under skin. Dorsal spines usually 4. . . *Apeltes quadracus*

     KK. Pelvic bones joined, forming median plate across belly behind pelvics. Dorsal spines usually more than 4

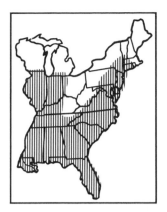

A. sayanus
(Gilliams)
Pirateperch

C. cornutus
Agassiz
Ricefish

C. agassizi
Putnam
Springfish

T. subterraneus
Girard
Southern Cavefish

A. spelaea
DeKay
Northern Cavefish

A. quadracus
(Mitchill)
Fourspine stickleback

    L.   Caudal peduncle without lateral keel. Dorsal spines usually 5. . . . . *Culea inconstans*

    LL. With keel. Spines usually 7-11 . . . . *Pungitius pungitius*

  JJ. Gill membranes not joined across isthmus. Usually 3 spines. . . . . *Gasterosteus aculeatus*

II. Dorsal spines connected by membranes or absent. Not free

  J.   Single dorsal fin with not more than one spine

    K.   Head scaleless

      L.   Pelvic axillary process present. Branchiostegal membranes free from isthmus

        M.   Dorsal fin base situated partly or entirely over anal fin base. Lateral line complete or well developed. Gill rakers few, short, and knoblike . . . . HIODONTIDAE, *HIODON*

        N.   Dorsal fin of 11 or 12 rays, its origin before anal. . *H. tergisus*

        NN. Dorsal fin 9 or 10 rays, its origin behind anal. . . *H. alosoides*

        MM. Dorsal base situated over pelvic base or slightly behind. No lateral line. Gill rakers many, long and slender . . CLUPEIDAE

          N.   Mouth terminal or inferior. Lower jaw included. Posterior dorsal fin ray greatly elongate. . . . . *DOROSOMA*

            O.   Mouth inferior. Anal rays 29-35. Lateral line scales more than 50 . . . . *D. cepedianum*

C. inconstans
(Kirtland)
Brook Stickleback

P. pungitius
(Linnaeus)
Ninespine Stickleback

G. aculeatus
Linnaeus
Threespine Stickleback

H. tergisus
LeSueur
Mooneye

H. alosoides
(Rafinesque)
Goldeye

D. cepedianum
(LeSueur)
Gizzard Shad

OO. Mouth terminal. Anal rays 17-25. Lateral line scales fewer than 50 . . . . *D. petenensis*

NN. Mouth terminal. Jaws equal or the upper included. Posterior dorsal fin ray not elongate. . . . . *ALOSA*

O. Peritoneum black . . . . *A. aestivalis*

OO. Peritoneum silvery or pale

   P. Premaxillae meeting at an acute angle forming deep notch

     Q. No teeth on tongue or jaws

       R. Gill rakers long and slender, about 60 on lower arm of first arch. . . . . *A. sapidissima*

       RR. Gill rakers shorter, about 40 on this arm . . *A. alabamae*

     QQ. Tongue with small patch or row of teeth. Jaws with weak teeth except in adult

       R. Tongue with small patch of teeth. Black spot on back of opercle . . . . *A. pseudoharengus*

       RR. Tongue with single row of teeth, no black spot . . . . *A. ohiensis*

   PP. Premaxillae meeting in obtuse angle forming shallow notch

     Q. Tongue teeth sometimes present. Lower jaw teeth only in young. Sides with faint longitudinal stripes . . . . *A. mediocris*

D. petenensis
(Gunther)
Threadfin Shad

A. aestivalis
(Mitchill)
Glut Herring

A. sapidissima
(Wilson)
American Shad

A. alabamae
Jordan and Evermann
Alabama Shad

A. pseudoharengus
(Wilson)
Alewife

A. ohiensis
Evermann
Ohio Shad

A. mediocris
(Mitchill)
Hickory Shad

QQ. Tongue teeth present. Teeth on lower jaw in young and
usually in adults. Sides without stripes. . *A. chrysochloris*

LL. Pelvic Axillary process absent. Branchiostegal membranes
united to isthmus

M. Dorsal fin usually with less than 10 dorsal rays except in
carp and goldfish in which the mouth is not sucker-like. . .
. . . . CYPRINIDAE (see key pg. 40)

MM. Dorsal rays usually 10 or more; if only 9 then lateral line
absent or reduced. Mouth typically sucker-like . . . . . .
. . . . CATOSTOMIDAE (see key pg. 76)

KK. Head at least partly scaled. Cheeks partly or entirely scaled

L. Tail deeply forked. Occiput without scales. Ducklike snout.
. . . . ESOCIDAE, *ESOX*

M. Opercle completely scaled

N. Branchiostegals 14-16. Snout 2.2-2.4 in head. Scales
about 124 in lateral line. . . . . *E. niger*

NN. Branchiostegals 11-14. Snout 2.4-3.1 in head. Scales
104-114 in lateral line. . . . . *E. americanus*

MM. Opercles with lower half naked

N. Cheeks scaleless on lower half. Branchiostegals 17-19. .
. . . .*E. masquinongy*

NN. Cheeks fully scaled. Branchiostegals usually 14-16 . *E. lucius*

LL. Tail rounded. Occiput with large scales. Jaws short

M. Premaxillaries not protractile . . . . .UMBRIDAE, *UMBRA*

A. chrysochloris
(Rafinesque)
Skipjack Herring

E. niger
LeSueur
Chain Pickerel

E. americanus
Gmelin
Redfin Pickerel

E. masquinongy
Mitchill
Muskellunge

E. lucius
Linnaeus
Northern Pike

N. Body with about 14 vertical bars, but may be indistinct. Lower jaw light . . . . *U. limi*

NN. Body with about 12 longitudinal streaks. Lower jaw dark.
. . . . *U. pygmaea*

MM. Premaxillaries protractile

N. Anal fin of male similar to that of female. Third anal ray branched. . . . . CYPRINODONTIDAE (see key pg. 88)

NN. Anal fin of male unlike that of female, modified into intro-mittent organ. Third anal ray unbranched. Living young produced. . . . . .POECILIIDAE

O. Teeth in villiform bands. Lateral bands absent

P. Dorsal rays 12-16. Dorsal origin over or in front of anal origin . . . . *Mollienesia latipinna*

PP. Dorsal rays 8-10. Dorsal origin well behind anal origin
. . . . *Gambusia affinis*

OO. Teeth in single series. Lateral band present and crossed by 6-9 vertical bars . . . . *Heterandria formosa*

JJ. Two dorsal fins or dorsal fin with more than one spine

K. Spines in first dorsal frail and readily bent

L. Pelvic fins abdominal or subadominal. Pectoral fins high, usually above body axis . . . . ATHERINIDAE

M. Premaxillae viewed from above forming a pointed beak. Scales more than 50 in lateral series . . . . *Labidesthes sicculus*

MM. Premaxillae crescent shaped; not forming a pointed beak. Less than 50 scales in lateral line . . . . *MENIDIA*

U. limi
(Kirtland)
Central Mudminnow

U. pygmaea
(DeKay)
Eastern Mudminnow

M. latipinna
LeSueur
Sailfin Molly

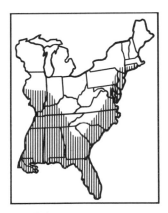

G. affinis
(Baird and Girard)
Gambusia

H. formosa
Agassiz
Least Killifish

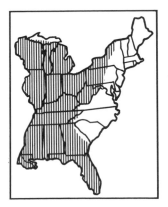

L. sicculus
(Cope)
Brook Silversides

       N.   Predorsal scales about 14-16. . . . . *M. beryllina*

      NN. Predorsal scales about 19-22. . . . . *M. extensa*

    LL. Pelvic fins thoracic or nearly so. Pectorals usually inserted below body axis

     M. Single anal spine present. .E LEOTRIDAE, *Dormitator maculatus*

     MM. Anal spine absent . . . . COTTIDAE

       N.   Gill membranes free from isthmus. Dorsals widely separated . . . . *Triglopsis thompsoni*

       NN. Gill membranes attached to side of wide isthmus. Dorsals joined, at least at base . . . . *COTTUS*

         O.   Preopercular spine long and spirally curved . . . . . . *C. ricei*

         OO. Spine, if present, not long and spirally curved

          P.   Lateral line complete or nearly so . . . . *C. carolinae*

          PP. Lateral line incomplete

           Q.   Palatine teeth absent

            R.   Usually 3 pelvic rays . . . . *C. cognatus*

            RR. Usually 4 pelvic rays . . . . *C. baileyi*

           QQ. Palatine teeth present. 4 pelvic rays . . . . *C. bairdi*

KK. Dorsal spines stiff and sharp

  L.   Anal spines 3 or more

    M.   Pseudobranchia well developed, exposed. Anal spines 3

      N.   Maxilla not sheathed by preorbital when mouth is closed. Jaws without incisorlike teeth . . . . SERRANIDAE

M. beryllina
(Cope)
Tidewater Silversides

M. extensa
Hubbs and Raney
Waccamaw Silversides
(Waccamaw Lake only)

D. maculatus
(Bloch)
Fat Sleeper

T. thompsoni
Girard
Deepwater Sculpin

C. ricei
Nelson
Spoonhead Sculpin

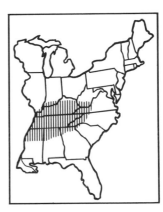

C. carolinae
(Gill)
Banded Sculpin

C. cognatus
Richardson
Slimy Sculpin

C. baileyi
Robbins
Black Sculpin

C. bairdi
Girard
Mottled Sculpin

O. Dorsal fins separate. Lower jaw projecting. Sides marked with bold stripes that are typically continuous. . . . . *Roccus chrysops*

OO. Dorsal fins joined by a membrane. Lower jaw not projecting. Sides marked with pale stripes or bold, discontinuous ones

  P. Longest dorsal spine about 1/2 head length. Sides marked with faint streaks. Dorsal fins well connected . . . . *R. americana*

  PP. Longest dorsal spine more than 1/2 head length. Sides marked with bold interrupted lines. Dorsal fins slightly connected. . . . . *R. mississippiensis*

NN. Maxilla sheathed for most of its length when mouth is closed. Incisor like teeth present. *Archosargus probatocephalus*

MM. Pseudobranchia, if present, small and concealed by a membrane. Anal spines 3 or more . . . . CENTRARCHIDAE (see key pg. 94)

LL. Anal spines 1 or 2

  M. Lateral line not extending well onto caudal fin. 2nd anal spine, when present, not long and stout . . . . PERCIDAE (see key pg. 102)

MM. Lateral line extended onto caudal fin. 2nd anal spine very long and stout . . . . SCIAENIDAE, *Aplodinotus grunniens*

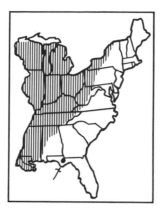

R. chrysops
(Rafinesque)
White Bass

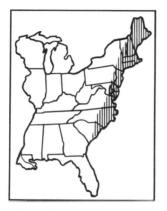

R. americana
(Gmelin)
White Perch

R. mississippiensis
Jordan and Eigenmann
Yellow Bass

A. probatocephalus
(Walbaum)
Sheepshead

A. grunniens
Rafinesque
Freshwater Drum

KEY TO SPECIES OF ICTALURIDAE

A. Adipose fin free at posterior end

  B. Premaxillary band of teeth with lateral backward processes forming a broad U with its open end backward. . . . . *Pylodictus olivaris*

  BB. Premaxillary band nearly straight. No posterior processes. . . . . *ICTALURUS*

    C. Caudal fin deeply forked

      D. Anal rays 18-22. Tail lobes not pointed, the upper lobe longer . . . . *I. catus*

      DD. Anal rays 23-26. Lobes sharp, about equal

        E. Anal rays 30-36. Tips of anal rays form a nearly straight line. . *I. furcatus*

        EE. Anal rays 23-30. Tips forming an arc . . . . *I. punctatus*

    CC. Caudal fin not deeply forked

      D. Anal rays 16-18 . . . . *I. platycephalus*

      DD. Anal rays 17-27

        E. Chin barbels white. Anal fin low anteriorly so that anterior rays are very little longer than those near posterior end. Posterior anal ray tips reach to or almost to caudal base. . . . . *I. natalis*

        EE. Chin barbels usually dark (but may be light if fish taken from muddy water). Anal rays near anterior end of fin much longer than those near posterior end

          F. Serrae on posterior edge of pectoral spines weak or absent. Fin membranes conspicuously black . . . . *I. melas*

          FF. Strong serrae. Membranes not blackened . . . . *I. nebulosus*

AA. Adipose fin adnate to back and caudal fin, no free posterior margin. . . . *NOTURUS*

P. olivaris
(Rafinesque)
Flathead Catfish

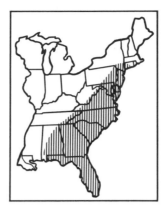

I. catus
(Linnaeus)
White Catfish

I. furcatus
(LeSueur)
Blue Catfish

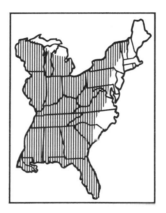

I. punctatus
(Rafinesque)
Channel Catfish

I. platycephalus
(Girard)
Flathead Bullhead

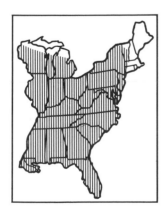

I. natalis
(LeSueur)
Yellow Bullhead

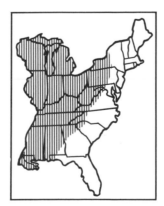

I. melas
(Rafinesque)
Black Bullhead

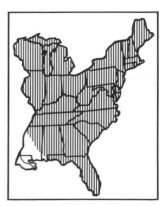

I. nebulosus
(LeSueur)
Brown Bullhead

B. Premaxillary band of teeth with lateral backward processes . . . . *N. flavus*

BB. No such backward processes

  C. Pectoral spines without serrae on posterior edges

    D. Pectoral spines without serrae on anterior edge. Sides with a narrow
black lateral streak. Jaws equal. *N. gyrinus*

    DD. Pectoral spines with anterior serrae. Color yellowish, somewhat mottled.
Lower jaw included . . . . *N. leptacanthus*

  CC. Pectoral spines with serrae on posterior edges (weak in *N. nocturnus*)

    D. 20 or more anal rays . . . . *N. funebris*

    DD. Less than 20 anal rays

      E. Adipose and caudal fins not separated by notch. Pectoral serrae limited
to a few near tip of spine on anterior edge and a few weak ones near base
on posterior edge . . . . *N. nocturnus*

    EE. Notch present. Pectoral serrae well developed on posterior edge

      F. Pectoral spines never with retrorse serrae; serae short, weak, erect
and sometimes bifurcate

        G. Adipose fin completely separated from caudal by notch. Pectoral
spine very short, its length contained 4 or 5 times in head. Tail
slightly emarginate. . . . . *N. gilberti*

        GG. Incomplete separation. Pectoral spine length contained less than 4
times in head. Tail rounded

        H. Jaws nearly equal. Fins usually not black edged. Head narrow and
flat . . . . *N. exilis*

        HH. Lower jaw included. Fins usually black edged. Head broad and
flat . . . . *N. insignis*

    FF. Serrae retrorse, stronger, and sometimes longer than width of spine

N. flavus
Rafinesque
Stonecat Madtom

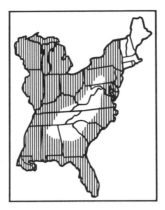

N. gyrinus
(Mitchill)
Tadpole Madtom

N. leptacanthus
Jordan
Gulf Coast Madtom

N. funebris
Jordan and Swain
Black Madtom

N. nocturnus
Jordan and Gilbert
Freckled Madtom

N. gilberti
Jordan and Evermann
Roanoke Madtom

N. exilis
Nelson
Slender Madtom

N. insignis
(Richardson)
Eastern Madtom

G.   Anterior pectoral serrae weak and less than 7. Posterior pectoral
     serrae length less than width of spine. Pectoral fin usually I, 9.
     Pelvic rays 8 . . . . *N. hildebrandi*

GG. Anterior pectoral serrae 14 to 33. Posterior pectoral serrae at least
     as wide as spine. Pectoral fin I, 8. Pelvic rays 9

  H.   Restricted to eastern North Carolina in Neuse, Tar and Little
       Rivers (similar to *N. miurus* but with much larger spines). . *N. furiosus*

  HH. Species of the Mississippi Valley

    I.   Dark blotch on adipose extending to fin margin. First dark saddle
         on back in front of dorsal fin, with little or no contact with dorsal
         soft rays . . . . *N. miurus*

    II. Dark blotch on adipose not extending to fin margin. First dark
         saddle on back mostly beneath dorsal fin . . . . *N. eleutherus*

N. hildebrandi
(Bailey and Taylor)
Least Madtom
Homochitto River only

N. furiosus
Jordan and Meek
Furious Madtom

N. miurus
Jordan
Brindled Madtom

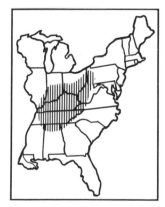

N. eleutherus
Jordan
Mountain Madtom

KEY TO SPECIES OF SALMONIDAE

A. Branchiostegals 13-19. 19-40 gill rakers on first arch. Anal fin with 13 or more rays . . . . *ONCORHYNCHUS*

  B. Gill rakers 30-40 on first arch . . . . *O. nerka*

  BB. Gill rakers 19-28 on first arch . . . . *O. kisutch*

AA. Branchiostegals less than 13. Gill rakers on first arch 20 or less. Anal fin with 13 or less rays

  B. Dorsal pattern with few to many dark spots on a lighter background. Vomer flattened . . . . *SALMO*

    C. Black or brown spots larger and more diffuse, scarcely developed on caudal fin

      D. Usually 11-13 scales in row from rear of adipose to but not including lateral line scale. Teeth on shaft of vomer weak or absent except in young. . . . . *S. salar*

      DD. Scales from adipose to lateral line usually 13-16. Teeth on shaft of vomer well developed. . . . . *S. trutta*

    CC. Black or brown spots small, sharp, numerous, typically well developed on caudal fin. Adipose fin of young with blackish margin and/or spots . . . .
. . . . *S. gairdneri*

  BB. Dorsal pattern of light spots on darker background. Vomer boat shaped . .
. . . . *SALVELINUS*

    C. Caudal fin deeply forked. Pyloric caeca 95-170. No bright colors . . . .
. . . . *S. namaycush*

    CC. Caudal fin little forked. Pyloric caeca less than 65. Red or orange spots on sides

      D. Dorsal rays 9. Anal rays usually 8. Maxilla reaching to middle of eye .
. . . . *S. aureolus*

      DD. Dorsal rays 10 or 11. Anal rays 9 or 10. Maxilla reaching at least to posterior margin of eye

        E. Back with conspicuous wormlike markings . . . . *S. fontinalis*

        EE. Back without such markings . . . . *S. oquassa*

O. nerka
(Walbaum)
Sockeye Salmon
(Introduced)

O. kisutch
(Walbaum)
Coho Salmon
(Introduced)

S. salar
Linnaeus
Atlantic Salmon

S. trutta
Linnaeus
Brown Trout

S. gairdneri
Richardson
Rainbow Trout

S. namaycush
(Walbaum)
Lake Trout

S. aureolus
Bean
Sunapee Trout

S. fontinalis
(Mitchill)
Brook Trout

S. oquassa
(Girard)
Blueback Trout

KEY TO SPECIES OF COREGONIDAE

A. 1 flap between nostrils. Gill rakers 20 or fewer . . . . . *Prosopium cylindraceum*

AA. 2 flaps between nostrils. Gill rakers more than 20 . . . . *COREGONUS*

  B. Premaxillae wider than long. Anterior edge of upper jaw directed downward
      and backward. Gill rakers on first arch fewer than 32 . . . . . *C. clupeaformis*

  BB. Premaxillae longer than wide. Edge of upper jaw directed forward and slightly
      downward. Gill arches more than 31

    C. Body deepest anteriad to its middle

      D. Gill rakers usually fewer than 33. . . . .*C. johannae*

      DD. Gill rakers usually more than 33

        E. Body small, compressed. Mandible thin and knobbed at tip. . . . . . *C. kiyi*

        EE. Body large, thick. Mandible thick, no knob . . . .*C. nigripinnis*

    CC. Body deepest near the middle

      D. Gill rakers usually 51 to 59. . . . .*C. hubbsi*

      DD. Gill rakers usually 34 to 52

        E. Gill rakers usually 43 to 52. . . . . *C. artedii*

        EE. Gill rakers usually 30 to 43

          F. Small species. Mandible thin with knob at tip . . . . *C. hoyi*

          FF. Larger species. The thick mandible usually lacking a knob

            G. Body little compressed. Lower jaw blackish, included . . . *C. reighardi*

            GG. Body usually more compressed. Lower jaw not blackish, equal or
                projecting

P. cylindraceum
(Pallas)
Round Whitefish

C. clupeaformis
(Mitchill)
Lake Whitefish

C. johannae
(Wagner)
Deepwater Chub

C. kiyi
(Koelz)
Kiyi

C. nigripinnis
(Gill)
Blackfin

C. hubbsi
(Koelz)
Ives Lake Cisco
(Ives Lake only)

C. artedii
LeSueur
Cisco

C. hoyi
(Gill)
Great Lakes Bloater

C. reighardi
(Koelz)
Shortnose Chub

H. Jaws usually equal . . . . *C. zenithicus*

HH. Upper jaw included

   I. Lateral line usually of 71-77 scales . . . . *C. bartletti*

   II. 78 to 85 scales in lateral line . . . . *C. alpenae*

C. zenithicus
(Jordan and Evermann)
Shortjaw Chub

C. bartletti
(Koelz)
Siskiwit Lake Cisco
Siskiwit Lake, Isle
Royal only

C. alpenae
(Koelz)
Longjaw Chub

KEY TO SPECIES OF CYPRINIDAE

A. Dorsal fin with more than 15 soft rays. Dorsal and anal fins with a strong spine-like ray

  B. Upper jaw with two barbels on each side. Lateral line usually with more than 32 scales. . . . . *Cyprinus carpio*

  BB. No barbels. Fewer than 30 lateral line scales . . . . *Carassius auratus*

AA. Dorsal fin with less than 10 rays. Dorsal and anal fin without a spine-like ray

  B. Abdomen behind pelvic fins with a fleshy keel. . . . . . *Notemigonus crysoleucas*

  BB. Abdomen behind pelvic fins rounded and scaled. No fleshy keel

    C. Female with an ovipositor often over twice as long as longest anal ray (introduced in Westchester Co., N.Y.) . . . . *Rhodeus amarus*

    CC. No ovipositor

      D. Barbels present, either at posterior edge of maxillary or in the groove above the maxillary well ahead of its posterior end

       E. Barbel in groove above maxillary well in advance of its posterior end

         F. Premaxillaries not protractile. Fleshy lobes of lower jaw covering only the posterior two-thirds of each side of jaw . . . *Parexoglossum laurae*

         FF. Premaxillaries protractile. Lower jaw normal . . . . SEMOTILUS

           G. Dorsal origin over pelvic insertion. About 45 scales in lateral line. . . . . *S. corporalis*

           GG. Dorsal origin behind pelvic insertion. Scales 50-75

             H. Black spot at dorsal origin, mouth reaching front of eye . . . . *S. atromaculatus*

             HH. No spot, mouth not reaching front of eye . . . . *S. margarita*

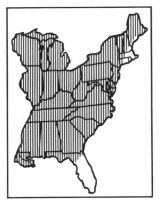

C. carpio
Linnaeus
Carp

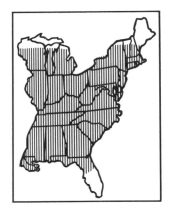

C. auratus
(Linnaeus)
Goldfish

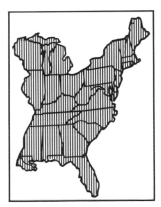

N. crysoleucas
(Mitchill)
Golden Shiner

R. amarus
(Bloch)
Bitterling
(Introduced,
Weschester Co.)

P. laurae
Hubbs
Tonguetied Chub

S. corporalis
(Mitchill)
Fallfish

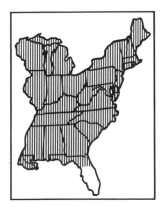

S. atromaculatus
(Mitchill)
Creek Chub

S. margarita
(Cope)
Pearl Dace

EE. Barbel at posterior end of maxillary

F.   Upper jaw not protractile . . . . *RHINICHTHYS*

  G.   Head narrow with fleshy snout overhanging inferior mouth . .*R. cataractae*

  GG. Head broader and shorter, snout not greatly overhanging the nearly
       terminal mouth . . . . *R. atratulus*

FF. Upper jaw protractile . . . . *HYBOPSIS*

  G.   Pharyngeal teeth 2, 4-4, 2 . . . .*H. plumbea*

  GG. Pharyngeal teeth 1, 4-4, 0 or 0, 4-4, 0

    H.   Mouth terminal or nearly so. Snout usually not extending beyond
        upper lip

      I.   Pharyngeal teeth 1, 4-4, 1 or 1, 4-4, 0. Large round caudal spot .
       . . . . *H. biguttata*

      II.   Teeth 0, 4-4, 0. Caudal spot small and round or diffuse and in-
        distinct

        J.   Breeding males with 7-10 large tubercles . . . . *H. bellica*

        JJ. With about 40 small tubercles

          K.   Snout contained about 8 times in standard length. Distance
            from dorsal origin to hypural equal to distance from dorsal
            origin to nostril . . . .*H. micropogon*

          KK. Snout about 9 in standard length. Distance from dorsal origin
            to hypural equal to distance from dorsal origin to eye . . . . .
            . . . . *H. leptocephala*

R. cataractae
(Valenciennes)
Longnose Dace

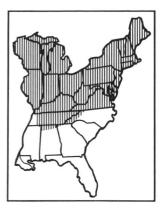

R. atratulus
(Hermann)
Blacknose Dace

H. plumbea
(Agassiz)
Northern Chub

H. biguttata
(Kirtland)
Hornyhead Chub

H. bellica
(Girard)
Southern Chub

H. micropogon
(Cope)
River Chub

H. leptocephala
(Girard)
Carolina Chub

HH. Mouth usually subterminal or inferior. Snout usually protruding beyond upper lip

I. Pharyngeal teeth 1, 4-4, 1 or 1, 4-4, 0

J. Mouth large, lower lip thick and papillose on its inner edge. Black spot at origin and another at end of dorsal base . . . *H. labrosa*

JJ. Mouth small, lips thin and not papillose. No black spots at dorsal base

K. No lateral band. Scales about 42 in lateral line . . . *H. storeriana*

KK. Dark or dusky lateral band present. 35-40 lateral line scales

L. Anal rays 7

M. About 38 lateral line scales, and about 16 before dorsal fin. No caudal spot . . . . *H. amblops*

MM. About 36 lateral line scales, and about 13 before dorsal fin. Caudal spot. . . . . *H. rubrifrons*

LL. Anal rays 8 . . . . *H. hypsinota*

II. Pharyngeal teeth 0, 4-4, 0

J. Lateral line scales fewer than 39

K. Body with scattered small spots. No caudal spot. Frontal area of head dark dorsally. Mouth ventral. . . . . *H. aestivalis*

KK. Lacking such spots. Caudal spot present. Frontal area with median light streak. Mouth slightly oblique . . . . *H. harperi*

J. Scales 39 to 62

H. labrosa
(Cope)
Thicklip Chub

H. storeriana
(Kirtland)
Silver Chub

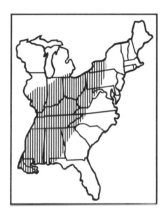

H. amblops
(Rafinesque)
Bigeye Chub

H. rubrifrons
(Jordan)
Redface Chub

H. hypsinota
(Cope)
Highback Chub

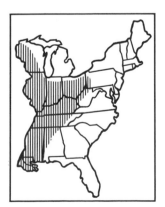

H. aestivalis
(Girard)
Speckled Chub

H. harperi
(Fowler)
Redeye Chub

    K.   Anal rays 9. Dorsal scales keeled . . . . *H. gelida*

  KK. Anal rays 7 or 8. No keels

      L.   Lower caudal lobe with black pigment above white edge. . *H. meeki*

    LL. Lobe uniformly pigmented

        M.   Lateral line scales 52-62. No dark blotches or speckles on body. Width between lower ends of gill openings contained more than six times in head . . . . *H. monacha*

      MM. Lateral line scales less than 52. Body with black blotches. Width contained fewer than six times in head

          N.   Body with irregular x-shaped marks, or with "V" marks on posterior myomeres

            O.   Body with "x" markings . . . . *H. x-punctata*

           OO. Posterior part of body with "V" markings . . . . . . *H. cahni*

        NN. Body with lateral and middorsal row of spots or blotches. No "x" or "V" marks

            O.   Row of lateral oval blotches. Eye diameter contained 3.5 to 4.0 in head . . . . *H. dissimilis*

           OO. Row of lateral rectangular blotches. Eye 2.9 to 3.5 in head . . . . *H. insignis*

DD. No barbels

  E. Cartilaginous ridge of lower jaw prominent and separated by a definite groove from lower lip. Mouth inferior. Intestine very long . . . . . .
    . . . . *Campostoma anomalum*

H. gelida
(Girard)
Roughscale Chub

H. meeki
Jordan and Evermann
Sicklefin Chub

H. monacha
(Cope)
Spotfin Chub

H. x-punctata
Hubbs and Crowe
Gravel Chub

H. cahni
Hubbs and Crowe
Slender Chub
(Clinch and Powell Rivers)

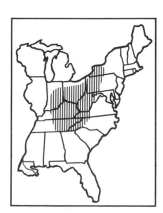

H. dissimilis
(Kirtland)
Streamline Chub

H. insignis
Hubbs and Crowe
Blotched Chub

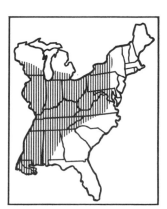

C. anomalum
(Rafinesque)
Stoneroller

EE. Cartilaginous ridge of lower jaw hardly evident. No such groove.
    Intestine and mouth variable

  F.   Lower lip obviously specialized

    G.   Lower lip with two fleshy lateral lobes. Premaxilla protractile. . .
. . . . *PHENACOBIUS*

      H.   Lateral line scales 42-53

        I.   Distinct round caudal spot . . . . *P. mirabilis*

        II. No distinct caudal spot . . . . *P. teretulus*

      HH. Lateral line scales 54 or more

        I.   Distinct caudal spot . . . . *P. uranops*

        II. Caudal spot vague and dusky

          J.   Usually 20 scales around caudal peduncle . . . . . . *P. crassilabrum*

          JJ. Scales around caudal peduncle 15-19 . . . . *P. catostomus*

    GG. Lower lip restricted to posterior 1/2 or 2/3 of jaw. Premaxilla not
        protractile . . . . *Exoglossum maxillingua*

  FF. Lower lip not highly specialized

    G.   Bones of the flattened lower surface of head with large externally
       visible cavernous chambers . . . . *Ericymba buccata*

    GG. Not as above

      H.   First dorsal ray more or less thickened, separated by a membrane
         from first well developed ray. Predorsal scales crowded and
         smaller than those on rest of body. Dark spot at front of dorsal
         fin, above base . . . . *PIMEPHALES*

        I.   Peritoneum silvery . . . . *P. vigilax*

        II. Peritoneum black

P. mirabilis
(Girard)
Suckermouth Minnow

P. teretulus
(Cope)
Kanawha Suckermouth Minnow

P. uranops
Cope
Stargazing Suckermouth Minnow

P. crassilabrum
Minckley and Craddock
Fatlips Minnow

P. catostomus
Jordan
Riffle Suckermouth Minnow

E. maxillingua
(LeSueur)
Cutlips Minnow

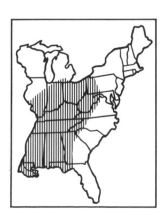

E. buccata
Cope
Silverjaw Minnow

P. vigilax
(Baird and Girard)
Bullhead Minnow

    J.   Mouth terminal. . . . . *P. promelas*

    JJ. Mouth ventral . . . . *P. notatus*

HH. First dorsal ray a thin splint attached to first well developed ray.
      Predorsal scales not greatly crowded nor smaller than those on
      rest of body. Anterior black spot if present is at very base of dorsal

  I.   48 or more lateral line scales

    J.   Lateral line incomplete . . . . *CHROSOMUS*

      K.   Lateral band single and dusky. Intestine short, S shaped . . .
          . . . . *C. neogaeus*

      KK. 2 lateral bands. Intestine elongate

        L.   Lateral bands not horizontal; lower one extending downward
            and backward from snout to anal base; upper beginning above
            anus and ending at caudal base . . . . *C. oreas*

        LL. Lateral bands horizontal and parallel

        M.   Distance from tip of snout to back of eye distinctly longer
            than rest of head. Mouth little oblique . . . . . *C. erythrogaster*

        MM. This distance equal to or little longer than rest of head.
            Mouth strongly oblique . . . . *C. eos*

    JJ. Lateral line complete or nearly so . . . . *CLINOSTOMUS*

      K.   65-70 scales in lateral line . . . . *C. elongatus*

      KK. 48-53 scales . . . . *C. funduloides*

  II.   Less than 48 lateral line scales

    J.   Intestine elongate, coiled and looped

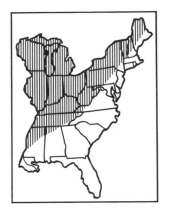

P. promelas
(Rafinesque)
Fathead Minnow

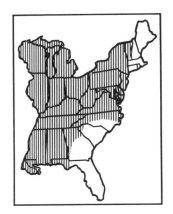

P. notatus
(Rafinesque)
Bluntnose Minnow

C. neogaeus
(Cope)
Finescale Dace

C. oreas
Cope
Mountain Redbelly Dace

C. erythrogaster
Rafinesque
Southern Redbelly Dace

C. eos
Cope
Mountain Redbelly Dace

C. elongatus
(Kirtland)
Redside Dace

C. funduloides
Girard
Rosy Dace

K.   Sides with conspicuous lateral band. Pharyngeal teeth hooked.
. . . . *Dionda nubila*

KK. Lateral band poorly developed or absent. Pharygeal teeth
lacking well developed hooks. . . . . *HYBOGNATHUS*

L.   Dorsal fin rounded. 1st developed dorsal ray shorter than
2nd or 3rd. . . . . *H. hankinsoni*

LL. Dorsal fin falcate. First developed dorsal ray equal to or
longer than 2nd or 3rd

M.   Eye nearly as long as snout, or longer . . . . *H. hayi*

MM. Eye about 2/3 the snout length

N.   Eye 4.0 to 5.5 in head. Width of head about equal to distance from posterior margin of orbit to snout tip . . *H. nuchalis*

NN. Eye 5.5 to 7.7 in head. Width of head greater than snout
to posterior orbit distance . . . . *H. placita*

JJ. Intestine short, s-shaped

K.   Pharyngeal teeth in main row 5-5. Mouth small and oblique,
and either lateral line very short (about 14 pores) or dorsal
with 9 rays

L.   Lateral line nearly complete. Pharyngeal teeth in single
row . . . . *Opsopoeodus emiliae*

LL. Lateral line incomplete. Two rows of pharyngeal teeth . . .
. . . . *Hemitremia flammea*

D. nubila
(Forbes)
Ozark Minnow

H. hankinsoni
Hubbs
Brassy Minnow

H. hayi
Jordan
Cypress Minnow

H. nuchalis
Agassiz
Silvery Minnow

H. placita
Girard
Plains Minnow

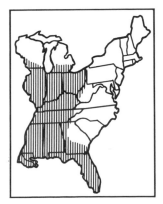

O. emiliae
Hay
Pugnose Minnow

H. flammea
(Jordan and Gilbert)
Flame Dace

KK. Not with above combination of characters. Pharynegeal teeth in main row 4-4. 8 dorsal rays. . . . . *NOTROPIS*

L. Posterior or middle interradial membranes of dorsal fin with distinct dark pigmented area, or anterior base of dorsal fin black

M. Black pigment on middle or posterior interradial membrane

N. Base of caudal fin creamy yellow (Pharyngeal teeth 1-4) . . . . *N. galacturus*

NN. Base of fin not prominently creamy yellow

O. Caudal spot distinct, either separate from lateral brand, or as widened lateral band

P. 9 or more anal rays

Q. Pharyngeal teeth 1-4, caudal spot elongate . *N. trichoistius*

QQ. Pharyngeal teeth 2-4. Caudal spot round . . . *N. zonistius*

PP. 7 or 8 anal rays. Dorsal fin with horizontal bar . *N. venustus*

OO. Caudal spot absent, or not distinct as described above

P. Pharyngeal teeth 0-4 or 2-4

Q. Anal rays 8 or 9

R. Pharyngeal teeth 0-4. Dorsal, anal and caudal fins white-tipped. . . . . *N. callisema*

RR. Teeth 2-4. Dark bar behind opercle. . . . *N. coccogenis*

QQ. Anal rays 10-11 (Pharyngeal teeth 2-4)

R. Lateral band extended as caudal spot . . *N. hypselopterus*

RR. Lateral band not extended . . . . *N. roseipinnis*

PP. Pharyngeal teeth 1-4

Q. Anal rays 8

R. Lateral band extends onto caudal base as "caudal spot"

S. Dorsal origin slightly behind pelvics and nearer snout tip than caudal base. Lateral line complete with 38 or 39 scales. Caudal spot in contact with lateral band. Middorsal stripe broad before dorsal and narrower behind. . . . .*N. callitaenia*

N. galacturus
(Cope)
Whitetail Shiner

N. trichoistius
(Jordan and Gilbert)
Tricolor Shiner

N. zonistius
(Jordan)
Bandfin Shiner

N. venustus
(Girard)
Blacktail Shiner

N. callisema
(Jordan)
Ocmulgee Shiner

N. coccogenis
(Cope)
Warpaint Shiner

N. hypselopterus
(Günther)
Sailfin Shiner

N. roseipinnis
Hay
Rosyfin Shiner

N. callitaenia
Bailey and Gibbs
Bluestripe Shiner

SS. Not with above combination of characters . *N. caeruleus*

RR. Lateral band not forming caudal spot at least in preserved specimens. (*N. niveus* has a faint spot when fresh)

  S. Mouth only slightly oblique. . . . .*N. niveus*

  SS. Mouth decidedly oblique

    T. Posterior margin of dorsal fin rounded   *N. chloristius*

    TT. Margin nearly straight . . . . *N. spilopterus*

QQ. Anal rays 9-11

  R. Anal rays usually 9

    S. Lateral line scales 38-40. Membranes between anterior dorsal rays with many small spots at least in large young and adults . . . .*N. whipplei*

    SS. Lateral line scales 34-35. Membranes lacking spots as described above . . . . *N. analostanus*

  RR. Anal rays usually 10

    S. Dorsal fin shorter than head. Lower jaw included. Faint black basicaudal spot . . . . *N. xaenurus*

    SS. Dorsal fin longer than head in breeding males. Jaws about equal. No caudal spot . . . . . . *N. pyrrhomelas*

MM. Intense black pigment at base of dorsal fin

  N. Scale rows crossing midline in front of dorsal fin fewer than 27. . . . . *N. matutinus*

  NN. 27 or more scale rows crossing midline

N. caeruleus
(Jordan)
Caerulean Shiner

N. niveus
(Cope)
Whitefin Shiner

N. chloristius
(Jordan and Brayton)
Greenfin Shiner

N. spilopterus
(Cope)
Spotfin Shiner

N. whipplei
(Girard)
Steelcolor Shiner

N. analostanus
(Girard)
Satinfin Shiner

N. xaenurus
(Jordan)
Altamaha Shiner

N. pyrrhomelas
(Cope)
Fieryblack Shiner

N. matutinus
(Cope)
Pinewoods Shiner

O. Lateral band very dark and crossing eye onto snout . . *N. lirus*

OO. Lateral band light anteriorly, and not crossing onto snout

    P. Anal rays usually 11-13. Body laterally flattened. Width of body usually contained 0. 9 to 1. 5 in head
. . . . *N. umbratilis*

    PP. Anal rays usually 10. Body more rounded, its width usually 1. 4 to 2. 1 in head . . . . *N. ardens*

LL. No dark pigment on dorsal fin as described above

  M. Pharyngeal teeth 0-4

    N. Anal rays typically 9

      O. Caudal base with conspicuous black spot. Eye goes about 3 times in head length . . . . *N. spectrunculus*

      OO. No spot. Eye about 3. 5 to 4. 0 in head . . . . . . *N. lutrensis*

    NN. Anal rays 7 or 8

      O. Anal rays typically 8

        P. Lateral-line scales with black crescents . . . *N. heterolepis*

        PP. Not with crescents, although scales may be marked with dots

          Q. Anterior lateral line scales much elevated

            R. Infraorbital canals complete . . . .*N. volucellus*

            RR. Infraorbital canals absent or poorly developed . . .
. . . . *N. buchanani*

          QQ. These scales not greatly elevated

N. lirus
(Jordan)
Mountain Shiner

N. umbratilis
(Girard)
Redfin Shiner

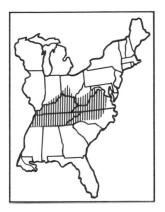

N. ardens
(Cope)
Rosefin Shiner

N. spectrunculus
(Cope)
Mirror Shiner

N. lutrensis
(Baird and Girard)
Red Shiner

N. heterolepis
Eigenmann and Eigenmann
Blacknose Shiner

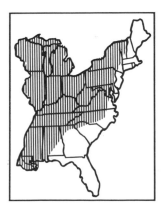

N. volucellus
(Cope)
Mimic Shiner

N. buchanani
Meek
Ghost Shiner

R. Lateral band black, extending forward to chin. Mouth nearly vertical . . . . *N. anogenus*

RR. Lateral band dusky. Mouth oblique. Caudal spot large and round . . . . *N. maculatus*

OO. Anal rays typically 7

P. Lateral band black and extending forward onto head

Q. Lateral line typically incomplete. Eye diameter greater than snout length. Lateral band on body wider than pupil . . . . *N. bifrenatus*

QQ. Lateral line complete or nearly so and characters otherwise not as above

R. Dark band completely ringing snout tip. Light band on snout above dark band. Caudal spot joined with lateral band. Upper lip dark. . . . . *N. alborus*

RR. Dark band on snout ends just in front of eye. No light band. Caudal spot much smaller than pupil and separate from lateral band. Upper lip tight. . . . . *N. procne*

PP. Lateral band silvery or dusky. Not extended forward onto head. Small wedge-shaped caudal spot

Q. Fish of Alabama river system . . . . *N. uranoscopus*

QQ. Fish north or west of the Alabama river system . . . . *N. stramineus*

MM. Pharyngeal teeth in two rows, 1-4 or 2-4

N. Pharyngeal teeth 1-4

O. Anal rays 9-11. Dark lateral band extending around snout, through eye, and extending posteriorly to caudal fin . . . . *N. cummingsae*

OO. Anal rays 7 or 8

N. anogenus
Forbes
Pugnose Shiner

N. maculatus
(Hay)
Tailight Shiner

N. bifrenatus
(Cope)
Bridled Shiner

N. alborus
Hubbs and Raney
Whitemouth Shiner

N. procne
(Cope)
Swallowtail Shiner

N. uranoscopus
Suttkus
Skygazer Shiner

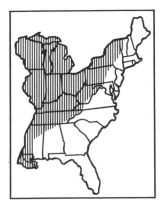

N. stramineus
(Cope)
Sand Shiner

N. cummingsae
Myers
Dusky Shiner

P. Anal rays usually 7. Mouth horizontal. Lateral band inconspicuous. . . . . *N. longirostris*

PP. Anal rays usually 8. Mouth and lateral band variable

Q. Posterior third of upper jaw concealed beneath suborbital. . . . . *N. amnis*

QQ. Upper jaw normal. No conspicuous extension behind angle of mouth

R. Very large and discrete black spot at caudal base. . . . . *N. callistius*

RR. Not as above

S. Peritoneum lacking black pigment. Lateral band not extending forward around head. Mouth nearly horizontal. Eye diameter equal to or less than snout length . . . . *N. dorsalis*

SS. Peritoneum silvery or black, but with much black pigment. Lateral band conspicuous and extending around snout. Mouth oblique. Eye diameter greater than snout length

T. Lateral line incomplete. Peritoneum silvery but with much black pigment . . . . *N. heterodon*

TT. Lateral line complete. Peritoneum black . . *N. boops*

NN. Pharyngeal teeth 2-4

O. Anal rays usually 7, 8 or 9

P. Anal rays usually 7 or 8

Q. Anal rays usually 7

R. Dark lateral band well developed

S. Anterior lateral line scales elevated. Mouth inferior. Predorsal dark streak obsolescent . . . . *N. asperifrons*

SS. Scales not elevated. Mouth terminal. Dark predorsal streak

N. longirostris
(Hay)
Longnose Shiner

N. amnis
Hubbs and Greene
Pallid Shiner

N. callistius
(Jordan)
Alabama Shiner

N. dorsalis
(Agassiz)
Bigmouth Shiner

N. heterodon
(Cope)
Blackchin Shiner

N. boops
Gilbert
Bigeye Shiner

N. asperifrons
Suttkus and Raney

T. Middorsal stripe dark and continuous around base
of dorsal fin to tail. Upper jaw longer than eye .
. . . .*N. baileyi*

TT. Middorsal stripe poorly developed or absent behind dorsal fin. Upper jaw shorter than or equal
to eye

U. Snout slightly overhanging mouth. Scales below
lateral line lacking pigment . . . . *N. petersoni*

UU. Snout not overhanging. Jaws equal, or the upper
shorter. Scales below lateral line with pigment

V. Scales below lateral line outlined with melanophores. . . . .*N. roseus*

V. Scales below lateral line pigmented in their
centers . . . . *N. xaenocephalus*

RR. Dark lateral band poorly developed, at least anteriorly

S. Black patches on body at bases of first 4 or 5 anal
rays . . . . *N. hypsilepis*

SS. Not as above. . . . . *N. blennius*

QQ. Anal rays usually 8

R. Lateral band usually very dark

S. Lateral line pores boldly outlined with elongate
melanophores. Light stripe above lateral band from
dorsal to caudal fins . . . . *N. leuciodus*

SS. Not as above

N. baileyi
Suttkus and Raney
Rough Shiner

N. petersoni
Fowler
Coastal Shiner

N. roseus
(Jordan)
Weed Shiner

N. xaenocephalus
(Jordan)
Coosa Shiner

N. hypsilepis
Suttkus and Raney
Cattahoochee Shiner

N. blennius
(Girard)
River Shiner

N. leuciodus
(Cope)
Tennessee Shiner

T. Lateral line incomplete. Lateral line scales
about 33. Interior of mouth, anal fin base and
caudal peduncle all with much black. . . *N. chalybaeus*

TT. Not with above combination of characters

U. Eye width less than snout length. Lateral band
on caudal peduncle as wide as pupil . . . *N. lutipinnis*

UU. Eye width equal to or less than snout length.
Lateral band on caudal peduncle not as wide as
snout length

V. Few or no melanophores below lateral line . .
. . . . *N. scabriceps*

W. Melanophores outlining first row of scales below lateral line . . . . *N. rubricroceus*

RR. Lateral band essentially silvery

S. Lateral line moderately to strongly decurved

T. Lateral line scales about 39 . . . . . *N. chlorocephalus*

TT. Lateral line scales about 36 . . . . *N. chiliticus*

SS. Lateral line straight or nearly so

T. Silvery fish, usually with pronounced caudal spot
(except in large adults). Mouth nearly horizontal
. . . . *N. hudsonicus*

N. chalybaeus
(Cope)
Ironcolor Shiner

N. lutipinnis
(Jordan and Brayton)
Yellowfin Shiner

N. scabriceps
(Cope)
Roughhead Shiner

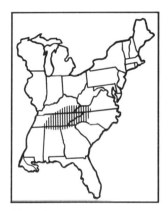

N. rubricroceus
(Cope)
Saffron Shiner

N. chlorocephalus
(Cope)
Greenhead Shiner

N. chiliticus
(Cope)
Redlip Shiner

N. hudsonicus
(Clinton)
Spottail Shiner

TT. Lateral line usually with rounded melanophores. Light band above lateral band from snout to caudal fin. Mouth oblique. . . . . *N. chrosomus*

PP. Anal rays usually 9

  Q. Sides of adults marked with conspicuous irregularly placed black cross blotches or with some scales (conspicuously higher than long) darkened with many small melanophores

    R. Sides with irregularly placed black cross blotches. Anterior lateral line scales not conspicuously elevated . . . . *N. cerasinus*

    RR. Not with cross blotches described above. Sides with irregularly blackened scales

      S. Predorsal scale count usually less than 22. . . . . *N. chrysocephalus*

      SS. Predorsal scales usually more than 24. . . . *N. cornutus*

  QQ. Sides not marked as above

    R. Breeding males with sides and fins silvery white. Snout white-tipped. Fishes of Atlantic coast *N. albeolus*

    RR. Lateral band silvery with little or no black pigment. Fishes of Mississippi drainage. . . . . *N. shumardi*

OO. Anal rays usually 10 or more (sometimes 9 in *N. altipinnis*)

N. chrosomus
(Jordan)
Colorful Shiner

N. cerasinus
(Cope)
Crescent Shiner

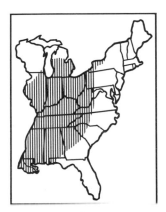

N. chrysocephalus
(Rafinesque)
Striped Shiner

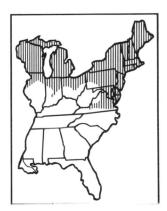

N. cornutus
(Mitchill)
Common Shiner

N. albeolus
Jordan
White Shiner

N. shumardi
(Girard)
Silverband Shiner

P.   Caudal base with evident spot . . . . *N. stilbius*

PP. Caudal base with vaque spot at most

  Q.   A pair of black crescents between nostrils . *N. photogenis*

  QQ. No crescents

    R.   Species of Atlantic Coast N. of Georgia (but see *N. rubellus* under RR, which also occurs on Atlantic Coast).

      S.   Dorsal origin nearer base of caudal fin than front of eye. Height of dorsal fin about one-half distance from dorsal origin to occiput . . . .*N. amoenus*

      SS. Dorsal origin nearer front of eye than base of caudal fin or about midway between these points. Height of dorsal fin more than one-half distance from dorsal origin to occiput

N. stilbius
(Jordan)
Silverstripe Shiner

N. photogenis
(Cope)
Silver Shiner

N. amoenus
(Abbott)
Attractive Shiner

T. Snout marked by a dark preorbital blotch which extends onto anterior half of lips and is bordered above by a light streak which passes through nostrils and around snout tip . . . . *N. altipinnis*

TT. Anterior snout darkened but light streak lacking

  U. Lateral line usually with 38 to 41 scales. Predorsal scales usually 19-23 . . . . . *N. semperasper*

  UU. Lateral line usually with 34 to 38 scales. Predorsal scales usually 14 to 19 . . . . . . *N. scepticus*

RR. Species of Great Lakes, Gulf Coast or Mississippi drainages

  S. Species restricted to Gulf Coast

    T. 26-31 rows of scales crossing midline; dorsal, pelvic and anal fins margined with black . . . *N. bellus*

    TT. 15-18 rows of scales crossing midline; fins not margined with black . . . . *N. signipinnis*

  SS. Species not so restricted

    T. Distance from dorsal origin to end of hypural plate equal to or greater than distance from dorsal origin to nostrils. Black dashes above and below lateral line pores . . . . *N. ariommus*

    TT. This distance less than dorsal origin to nostrils. Dashes present or absent

      U. Middorsal stripe rather broad and prominent before and behind dorsal fin. Lateral line pores often marked by melanophores, at least anteriorly.

N. altipinnis
(Cope)
Highfin Shiner

N. semperasper
Gilbert
Roughhead Shiner

N. scepticus
(Jordan and Gilbert)
Carolina Shiner

N. bellus
Hay
Pretty Shiner

N. signipinnis
Bailey and Suttkus
Flagfin Shiner

N. ariommus
(Cope)
Popeye Shiner

V.   Lateral band diffuse, but with about same intensity throughout its length. Chin with some black pigment. . . . . *N. percobromus*

VV. Lateral band intense black on caudal peduncle, but dimishing anteriorly. Chin without black pigment except on lips . . . . *N. rubellus*

UU. Middorsal stripe of fine lines or rows of dots before dorsal fin and faintly or well marked behind. Lateral line pores usually not marked by melanophores

V.   Predorsal scale rows 17-20. Scales of dorsum usually well outlined with melanophores. Middorsal stripe distinct behind dorsal fin . . . .
. . . . *N. atherinoides*

VV. Predorsal scale rows 23-28. Scales of dorsum rather evenly covered with melanophores. Middorsal stripe more diffuse forward . . . *N. fumeus*

N. percobromus
(Cope)
Plains Shiner

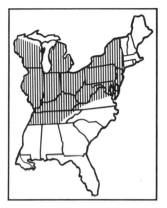

N. rubellus
(Agassiz)
Rosyface Shiner

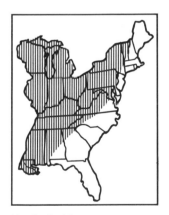

N. atherinoides
Rafinesque
Emerald Shiner

N. fumeus
Evermann
Ribbon Shiner

KEY TO SPECIES OF CATOSTOMIDAE

A. Dorsal fin with more than 20 rays

  B. Eye closer to posterior edge of opercular membrane than to tip of snout. Lateral line scales more than 50 . . . . *Cycleptus elongatus*

  BB. Eye closer to tip of snout than to opercular membrane. Lateral line scales fewer than 50

    C. Subopercle broadest at its middle, its edge forming an even arc. Eye round . . . . *ICTIOBUS*

      D. Mouth oblique and nearly terminal. Tip of upper lip about level with lower margin of eye. Upper jaw about as long as snout . . . . *I. cyprinellus*

      DD. Mouth nearly horizontal, subterminal. Tip of lower jaw below level of eye. Upper jaw shorter than snout

        E. Body depth about 2.5 in standard length. Head width more than 5 in standard length. Eye contained 2.0 to 2.5 in length of snout in specimens more than six inches long. . . . . *I. niger*

        EE. Body depth about 3 in standard length. Head width less than 5 in standard length. Eye contained 1.5 to 2.0 in snout length in specimens more than six inches long . . . . *I. bubalus*

    CC. Subopercle subtriangular, broadest below middle. Eye longer than high . . . . *CARPIODES*

      D. Scales 37 to 40 in lateral line. No knob-like projection of lower jaw at symphysis. Snout produced, tip of lower lip clearly in advance of anterior nostril; distance from tip of snout to anterior nostril equal or greater than eye diameter . . . . *C. cyprinus*

      DD. Scales 33-36. Knob present. Snout rounded, tip of lower jaw not much in advance or behind anterior nostril.

C. elongatus
(LeSueur)
Blue Sucker

I. cyprinellus
(Valenciennes)
Bigmouth Buffalofish

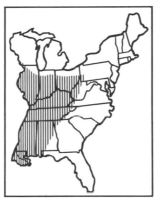

I. niger
(Rafinesque)
Black Buffalofish

I. bubalus
(Rafinesque)
Smallmouth Buffalofish

C. cyprinus
(LeSueur)
Quillback Carpsucker

E.   Anterior rays of dorsal fin greatly elevated, often extending length of fin (except in yg). Distance from tip of snout to anterior nostril contained more than 3 times in postorbital length of head . . . . *C. velifer*

EE. Rays not as elevated, shorter than dorsal base. Snout-nostril distance less than three times in postorbital length . . . . *C. carpio*

AA. Dorsal fin with less than 20 rays

  B.   Lateral line scales more than 55 . . . . *CATOSTOMUS*

    C.   Lateral line scales 55 to 85. Snout not projecting much beyond upper lip . . . . *C. commersoni*

    CC. Lateral line scales more than 85. Snout projecting beyond upper lip . . . . *C. catostomus*

  BB. Lateral line scales less than 55

    C.   Lateral line poorly developed or absent

      D.   Mouth horizontal, inferior. Lateral line poorly developed. Usually a black spot on each scale. . . . . *Minytrema melanops*

      DD. Mouth oblique. Lateral line absent. No scale spots . . . . *ERIMYZON*

        E.   Anal fin of male bilobed, not much longer than head in either sex. Dorsal fin rounded

          F.   Scales 34-38 . . . . *E. sucetta*

          FF. Scales 39-45 . . . . *E. oblongus*

        EE. Anal fin of male not bilobed, longer than head. Dorsal fin sharply pointed in half grown individuals. Species of Gulf Coast. . . . . *E. tenuis*

    CC. Lateral line well developed

      D.   Head concave between eyes . . . . *HYPENTELIUM*

C. velifer
(Rafinesque)
Highfin Carpsucker

C. carpio
(Rafinesque)
River Carpsucker

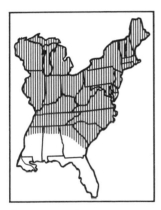

C. commersoni
(Lacepede)
White Sucker

C. catostomus
(Forster)
Longnose Sucker

M. melanops
(Rafinesque)
Spotted Sucker

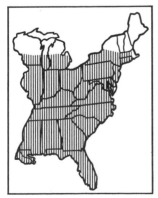

E. sucetta
(Lacepede)
Lake Chubsucker

E. oblongus
(Mitchill)
Creek Chubsucker

E. tenuis
(Agassiz)
Sharpfin Chubsucker

E.   Dorsal rays usually 10. Sides above with dark longitudinal stripes. Pelvics and pectorals red . . . . *H. etowanum*

EE. Dorsal rays usually 11. Longitudinal stripes present or absent. Fins light orange or brown

    F.   Lateral line scales usually more than 44. Saddle before dorsal well developed, crossing back. No light streaks on dorsal scale rows, larger size . . . . *H. nigricans*

    FF. Lateral line scales usually less than 44. Saddle before dorsal obsolescent, not crossing midline of back. Light streaks present dorsally, maximum size 6". . . . . *H. roanokense*

DD. Head rounded between eyes

    E.   Premaxillaries not protractile. Halves of lower lip separated . . . . *Lagochila lacera*

    EE. Premaxillaries protractile. Halves of lower lip joined at midline . . . . *MOXOSTOMA*

        F.   Air bladder of two chambers (often small)

           G.   Dorsal fin with large black blotch. Peritoneum silvery . . . *M. atripinne*

          GG. No such blotch. Peritoneum black.

           H.   Head rounded. Lower lip papillose. Vertical line from posterior edge of lower lip extends through eye. Greatest outside width of lips at least as wide as postocular length of head, and contained about 2.3 in standard length . . . . *M. hamiltoni*

          HH. Head bluntly pointed. Lower lips papillose behind, but becoming plicate anteriorly. Vertical line from posterior edge of lower lips passes in front of eye. Greatest outside width of lips less than postocular length of head and about 2.6 in standard length . *M. rhothoeca*

        FF. Air bladder of three chambers

H. etowanum
(Jordan)
Alabama Hogsucker

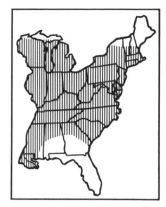

H. nigricans
(LeSueur)
Northern Hogsucker

H. roanokense
Raney and Lachner
Roanoke Hogsucker

Lagochila lacera
Jordan and Brayton
Harelip Sucker

M. atripinne
Bailey
Blackfin Sucker

M. hamiltoni
(Raney and Lachner)
Rustyside Sucker
(Roanoke only)

M. rhothoeca
(Thoburn)
Torrent Sucker

G. Scales around caudal peduncle 16

    H. Dorsal rays 13 or more. Dorsal margin convex

        I. Pharyngeal teeth large, bases of largest ones more than half the length of the arch. Middle of pupil nearer snout tip than posterior opercular margin. Nostril above posterior margin of upper lip . . . .*M. hubbsi*

        II. Bases of largest pharyngeal teeth less than half the length of the arch. Middle of pupil nearer snout tip than posterior opercular margin. Nostril above posterior margin of upper lip . *M. valenciennesi*

    HH. Dorsal rays 12 or less. Dorsal margin falcate

        I. Eye diameter greater than least distance from orbit to opercular edge. Anterior edge of eye above or slightly behind posterior edge of lower lip. Nostril far in front of hind margin of lower lip . . . . *M. ariommum*

        II. Eye diameter less than least distance from orbit to opercular edge. Anterior edge of eye and nostril well behind posterior edge of lip

          J. Tips of anterior dorsal rays black. Lateral line scales 43 or less. Eye length greater than 2/3 the outside width of mouth in adults. . . . . *M. cervinum*

          JJ. Anterior dorsal and caudal tips not black, dorsal sometimes with dusky edges. 44 or more lateral line scales. Eye length less than 2/3 the outside width of mouth in adults

M. hubbsi
Legendre
Copper Redhorse

M. valenciennesi
Jordan
Greater Redhorse

M. ariommum
Robins and Raney
Bigeye Jumprock

M. cervinum
(Cope)
Black Jumprock

K. Head depth at occiput contained 3 or more times in predorsal length. Middle lateral line scales much smaller than eye. Eye diameter contained more than 5 times in head length. Body depth at dorsal origin less than head length

    L. Head deeper than wide. Dorsal rays usually 12 . . . . *M. lachneri*

    LL. Head wider than deep. Dorsal rays usually 10 or 11 . . . . .
. . . . *M. rupiscartes*

KK. Head depth less than 3 times in predorsal length. Middle lateral line scales equal to or larger than eye. Eye diameter contained fewer than 5 times in head. Depth of body at dorsal origin equal to or greater than head length. . . . . *M. robustum*

GG. Scales around caudal peduncle 12

  H. Species of the Atlantic Coastal Plain South of the Saint Lawrence Drainage

    I. Lips papillose . . . . *M. papillosum*

    II. Lips essentially plicate . . . . *M. macrolepidotum*

  HH. Species of the Great Lakes Drainage area, West of the Appalachians or of the Gulf Coast

    I. Caudal fin white edged with a black longitudinal streak through lower lobe . . . . *M. poecilurum*

    II. Caudal fin with black pigment more evenly distributed

      J. Posterior edges of lower lips meeting in a straight line

M. lachneri
Robbins and Raney
Greater Jumprock

M. rupiscartes
Jordan and Jenkins
Striped Jumprock

M. robustum
(Cope)
Robust Redhorse

M. papillosum
(Cope)
Suckermouth Redhorse

M. macrolepidotum
(LeSueur)
Eastern Redhorse

M. poecilurum
Jordan
Blacktail Redhorse

K. Scales with basal spots. Pelvic, pectoral and caudal fins red. 9 or 10 pelvic rays

    L. Usually 9 pelvic rays. Anterior rays of depressed dorsal not extending to end of last ray of depressed dorsal. . . . *M. aureolum*

    LL. Usually 10 pelvic rays in one or both fins. Anterior rays extending past last ray of depressed dorsal . . . . *M. breviceps*

KK. Scales lacking basal spots. Fins not red. 10 pelvic rays . . .
. . . . *M. duquesnei*

JJ. Posterior edges of lower lips meeting in an obtuse V shaped line, or having a deep notch

    K. 15 or 16 dorsal rays, deep notch at junction of lower lips . . .
. . . . *M. anisurum*

    KK. 12-13 dorsal rays, posterior edges of lower lips meeting in obtuse V

        L. Scales with distinct dark spots at scale bases. Pharyngeal teeth few, molariform, not greatly compressed . . *M. carinatum*

        LL. Scales lacking dark spots. Pharyngeal teeth numerous, compressed . . . . *M. erythrurum*

(Three species of *Moxostoma, M. collapsum, M. coregonus* and *M. lachrymale,* are imperfectly known and could not be entered in the key. Maps are included for each.)

M. aureolum
(LeSueur)
Shorthead Redhorse

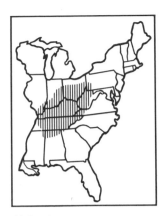

M. breviceps
(Cope)
Ohio Redhorse

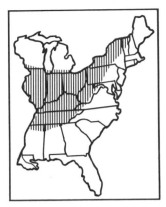

M. duquesnei
(LeSueur)
Black Redhorse

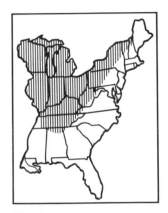

M. anisurum
(Rafinesque)
Silver Redhorse

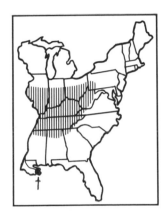

M. carinatum
(Cope)
River Redhorse

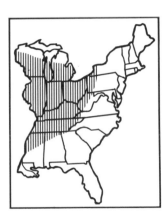

M. erythrurum
(Rafinesque)
Golden Redhorse

M. collapsum
(Cope)

M. coregonus
(Cope)

M. lachrymale
(Cope)

KEY TO SPECIES OF CYPRINODONTIDAE

A.   Jaw teeth unicuspid

  B.   Jaw teeth in single series

    C.   Lateral band from head to tail ₒ ₒ . . . . *Chriopeops goodei*

    CC. No lateral band, at least anteriorly

      D.   Body depth contained more than 4 times in standard length. A spot on sides or on caudal peduncle, or both . . . . *Leptolucania ommata*

      DD. Body depth contained fewer than 4 times in standard length. No spots or bars . . . . ₒ . . . . *Lucania parva*

  BB. Jaw teeth in more than one series

    C.   Fewer than 30 scales in lateral series. No pores on mandible. . .*Adinia xenica*

    CC. More than 30 scales in lateral series. 3-6 pores on mandibles . . .*FUNDULUS*

      D.   Usually more than 10 dorsal rays. Dorsal origin above, in front of or only slightly behind anal origin

        E.   Dorsal origin in front of anal origin

          F.   31-38 scales in lateral line

            G.   Mandibular lateral line pores 4 on each side . . . . *F. heteroclitus*

            GG. Pores 5 on each side . . . . *F. grandis*

          F.   41-60 scales in lateral line

            G.   Anal rays usually 10-12. Dorsal rays 13-15

              H.   Depth of caudal peduncle 3.2 to 3.6 in its length. 54-64 lateral line scales . . . . *F. waccamensis*

              HH. Depth 2.2 to 2.8 in its length. 41-50 lateral line scales . . *F. diaphanus*

            GG. Anal rays usually 13 or 14. Dorsal rays about 17 . . . . . . *F. seminolis*

C. goodei
Fowler
Redtail Killifish

L. ommata
(Jordan)
Ocellated Killifish

L. parva
(Baird and Girard)
Rainwater Fish

A. xenica
(Jordan and Gilbert)
Diamond Killifish

F. heteroclitus
(Linnaeus)
Mummichog

F. grandis
Baird and Girard
Gulf Killifish

F. waccamensis
Hubbs and Raney
Waccamaw Killifish
(Waccamaw Lake only)

F. diaphanus
(LeSueur)
Banded Killifish

F. seminolis
Girard
Seminole Killifish

EE. Dorsal origin over or a little behind anal origin

   F.   13-16 dorsal and anal rays. 50-60 lateral line scales

     G.   Branchiostegals 5. Dorsal rays usually 14 . . . . *F. catenatus*

     GG. Branchiostegals 4. Dorsal rays usually 13 . . . . *F. stellifer*

  FF. 10-11 dorsal and anal rays. 36-42 lateral line scales

     G.   Longitudinal stripes along scale rows. Usually more than 40 scales in lateral line

       H.   With about 14 vertical dark bars and a longitudinal streak along each scale row . . . . *F. confluentus*

       HH. Without vertical bars but with whitish or black streaks along scale rows . . . . *F. albolineatus*

     GG. With scattered dots and blotches of irregular size and arrangement, sometimes forming longitudinally elongate blotches. Lateral line scales usually less than 40

       H.   Fins with small brownish dots . . . . *F. pulvereus*

       HH. Fins plain, or speckled at base only . . . . *F. rathbuni*

DD. Usually no more than 10 dorsal rays. Dorsal origin distinctly behind anal origin

   E.   No conspicuous markings except 2 rows of spots on sides, sometimes coalesced to form short and indistinct bars . . . . *F. jenkinsi*

  EE. No such spots, but with other distinct markings

     F.   Single black lateral band from snout to caudal base

       G.   Lateral band intense and rather even edged . . . . *F. olivaceous*

       GG. Lateral band less intense and tending to form crossbars . . . . *F. notatus*

     FF. No single black band

       G.   No horizontal streaks, but vertical bars present. No subocular bar

F. catenatus
(Storer)
Northern Studfish

F. stellifer
Jordan and Meek
Southern Studfish

F. confluentus
Goode and Bean
Marsh Killifish

F. albolineatus
Gilbert
Whiteline Topminnow

F. pulvereus
(Evermann)
Bayou Killifish

F. rathbuni
Jordan and Meek
Speckled Killifish

F. jenkinsi
(Evermann)
Saltmarsh Topminnow

F. olivaceous
(Storer)
Blackspotted Topminnow

F. notatus
(Rafinesque)
Blackstripe Topminnow

       H.   Males with jet black spot on dorsal fin . . . . *F. luciae*

      HH. No such spot

          I.   Anal rays about 11 . . . . *F. chrysotus*

          II. Anal rays about 8-11 . . . . *F. cingulatus*

    GG. Sides with horizontal streaks. Vertical bars present or absent. Sub-ocular bar present

       H.   About 12 scales around caudal peduncle . . . . *F. notti*

       HH. About 16 scales around caudal peduncle . . . . *F. lineolatus*

AA. Jaw teeth tricuspid

  B.   Dorsal fin with 16 to 18 rays. . . . . *Jordanella floridae*

  BB. Dorsal fin with 9 to 12 rays . . . . *Cyprinodon variegatus*

F. luciae
(Baird)
Spotfin Killifish

F. chrysotus
Holbrook
Golden Topminnow

F. cingulatus
Valenciennes
Banded Topminnow

F. notti
(Agassiz)
Starhead Topminnow

F. lineolatus
(Agassiz)

J. floridae
Goode and Bean
Flagfish

C. variegatus
Lacepede
Variegated Pupfish

KEY TO SPECIES OF CENTRARCHIDAE

A. Lateral line absent. Branchiostegals 5. Very small, usually under 50 mm .
. . . . *ELLASSOMA*

  B. 38-45 lateral line scales . . . . *E. zonatum*

  BB. 27-30 lateral line scales . . . . *E. evergladei*

AA. Lateral line present (may be incomplete). Branchiostegals 6 or 7. Adults greater than 50 mm

  B. Anal spines typically 3. Branchiostegals 6

   C. Lateral line scales 55-81. Body elongate . . . . *MICROPTERUS*

    D. Dorsal fin nearly divided with shortest spine in notch less than half the length of the longest dorsal spine . . . . *M. salmoides*

    DD. Dorsal fins more broadly connected, shortest spine in notch more than half the length of longest spine

     E. Scale rows above lateral line 12 or 13, 20-23 below . . . . *M. dolomieui*

     EE. Scale rows above lateral line 7 to 10, 14 to 19 below

      F. Broken lateral band of often confluent blotches, not vertically elongate. 24 or 25 scale rows around caudal peduncle . . . . *M. punctulatus*

      FF. Vertical bars or blotches. Usually 26 to 30 scale rows around caudal peduncle

       G. Prominent spot at base of tail. Ventrolateral streaks poorly developed . . . . *M. notius*

       GG. No such spot. Ventrolateral streaks well developed . . . . . . *M. coosae*

  CC. Scales 26 to 54. Body deep

   D. Caudal fin emarginate.

E. zonatum
Jordan
Banded Pigmy Sunfish

E. evergladei
Jordan
Everglades Pigmy Sunfish

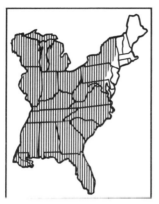

M. salmoides
(Lacepêde)
Largemouth Bass

M. dolomieui
Lacepêde
Smallmouth Bass

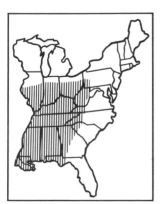

M. punctulatus
(Rafinesque)
Spotted Bass

M. notius
Bailey and Hubbs
Suwannee Bass

M. coosae
Hubbs and Bailey
Redeye Bass

E.   Teeth on tongue. Supramaxilla longer than width of maxilla . . . . . .
. . . . *Chaenobryttus gulosus*

EE. No teeth on tongue. Supramaxilla shorter than maxilla width . . . . *LEPOMIS*

  F.   Gill rakers long, much longer than wide, reaching base of 2nd or 3rd raker below

    G.   Mouth large, with posterior end of upper jaw extending well past anterior edge of eye. Pectoral fin short; reaches about to anterior edge of eye

      H.   Opercular bone long and stiff, and fractures if bent sharply forward

        I.   Lateral line complete, lateral line scales 45-53 . . . . . . *L. cyanellus*

        II. Lateral line incomplete, lateral line scales 31-40 . . . *L. symmetricus*

      HH. Short opercular bone, not fracturing when opercle is bent forward.
. . . . *L. humilis*

    GG. Small mouth, long pointed fin . . . . *L. macrochirus*

  FF. Gill rakers short, nearly as wide as long, usually not reaching base of second raker below

    G.   Opercular bone rather stiff to its margin, not fimbriate on its posterior edge

      H.   Pectoral fins short and rounded, when bent forward reaching about to anterior edge of eye . . . . *L. punctatus*

      HH. Pectoral fins long and pointed, extending well past anterior edge of eye

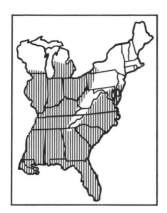

C. gulosus
(Cuvier)
Warmouth Sunfish

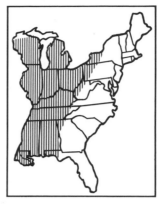

L. cyanellus
Rafinesque
Green Sunfish

L. symmetricus
Forbes
Bantam Sunfish

L. humilis
(Girard)
Orangespotted Sunfish

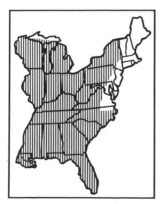

L. macrochirus
Rafinesque
Bluegill Sunfish

L. punctatus
(Valenciennes)
Spotted Sunfish

I.   No distinct spotting on soft dorsal. Opercular bone less stiff; it can be bent moderately forward without fracturing . . . *L. microlophus*

II.   Distinct spotting on soft dorsal. Opercular bone stiff. . .   *L. gibbosus*

GG. Opercular bone extended backward as thin flexible flap that can be bent sharply forward without breaking. Opercle usually fimbriate posteriorly

  H.   Opercle long and narrow, its margin deeply fimbriate. Opercular membrane dark to its margin. Gill rakers short, but not mere knobs . . . . *L. auritus*

HH. Opercle shorter and broader, its margin shallowly fringed in adult. Opercular membrane with lighter margin (greenish, red or white). Gill rakers mere knobs

    I.   Pectoral rays usually 13 to 15. Usually 5 or more rows of cheek scales . . . . *L. megalotis*

    II.   Pectoral rays usually 12. 4 rows of cheek scales. . . . *L. marginatus*

DD. Caudal fin rounded

  E.   10 dorsal spines. Black streak at front of dorsal fin . . . . *Mesogonistius chaetodon*

EE. Usually 9 dorsal spines. No black streak . . . . *ENNEACANTHUS*

  F.   Body with broad vertical bands. Opercular spot larger than pupil . . . . *E. obesus*

FF. Body with longitudinal light stripes. Opercular spot smaller than pupil . . . . *E. gloriosus*

BB. Anal spines rarely fewer than 5. Branchiostegals 7 (6 in *Amblopites*)

  C.   Gill rakers on first arch fewer than 15. Preopercle entire or weakly serrate

L. microlophus
(Günther)
Redear Sunfish

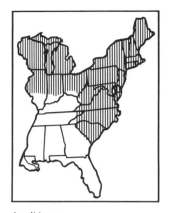

L. gibbosus
(Linnaeus)
Pumpkinseed Sunfish

L. auritus
(Linnaeus)
Yellowbelly Sunfish

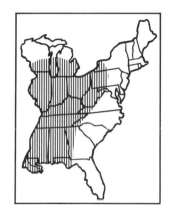

L. megalotis
(Rafinesque)
Longear Sunfish

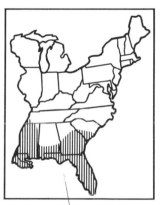

L. marginatus
(Holbrook)
Dollar Sunfish

M. chaetodon
(Baird)
Blackbanded Sunfish

E. obesus
(Girard)
Banded Sunfish

E. gloriosus
(Holbrook)
Bluespot Sunfish

    D.   Caudal fin emarginate. Ctenoid scales . . . . *AMBLOPLYTES*

        E.   Cheeks incompletely scaled, 10-12 rows of scales above lateral line . .
        . . . . *A. cavifrons*

      EE. Cheeks completely scaled, 7-9 rows of scales . . . . *A. rupestris*

   DD. Caudal fin rounded. Scales cycloid . . . . *Acantharchus pomotis*

CC. Gill rakers on first arch 25 or more. Preopercle finely serrate

    D.   Dorsal spines 5-8 . . . . *POMOXIS*

        E.   Dorsal spines 7 or 8. Dorsal base equal to or greater than distance
           from dorsal origin to posterior edge of eye . . . . *P. nigromaculatus*

      EE. Dorsal spines 6. Dorsal base less than this distance . . . . . . *P. annularis*

   DD. Dorsal spines 11-13 . . . . *Centrarchus macropterus*

A. cavifrons
Cope
Eastern Rockbass

A. rupestris
(Rafinesque)
Rockbass

A. pomotis
(Baird)
Mud Sunfish

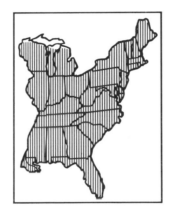

P. nigromaculatus
LeSueur
Black Crappie

P. annularis
Rafinesque
White Crappie

C. macropterus
(Lacepede)
Flier

KEY TO SPECIES OF PERCIDAE

A.   Preopercle strongly serrate. Maxilla extending at least to below middle of eye. Branchiostegals usually 7

   B.   Canine teeth absent. Pelvic fins close together. Body with dark vertical bands. Pseudobranchiae rudimentary . . . . *Perca flavesceus*

   BB. Canine teeth present. Pelvic fins separated by a distance about equalling their bases. Body with indistinct bands or saddles. Pseudobranchiae well developed . . . . *STIZOSTEDION*

      C.   Dorsal fins with rows of round, black spots (not in young). No prominent black blotch on posterior end of spinous dorsal. Dorsal soft rays 17-20. .
. . . . *S. canadense*

      CC. Indistinct dusky markings on dorsal. Large black blotch or posterior spinous dorsal. Rays 19-22 . . . . *S. vitreum*

AA. Preopercle entire or only weakly serrate. Maxillae usually not reaching middle of eye. Branchiostegals usually 6

   B.   Midline of belly usually with a row of enlarged and modified scales (may be inconspicuous). If such scales are absent, then a narrow row of scales just in front of the anus, with the front part of the belly naked . . . . *PERCINA*

      C.   Snout conical and fleshy, protruding over ventral mouth

         D.   Depth of body contained less than 5 times in standard length . . . . . . *P. rex*

         DD. Depth more than 5 times in standard length . . . . *P. caprodes*

      CC. Snout not conical and fleshy, mouth terminal

         D.   Anal fin of adult males extremely long, posterior rays extending beyond posterior end of depressed soft dorsal in large specimens

            E.   Spinous dorsal with conspicuous black spot on its posterior rays and a smaller one on anterior rays. Sides with indistinct blotches . . *P. shumardi*

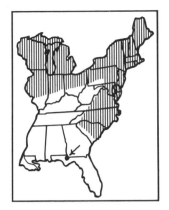

P. flavescens
(Mitchill)
Yellow Perch

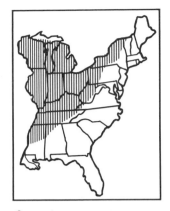

S. canadense
(Smith)
Sauger

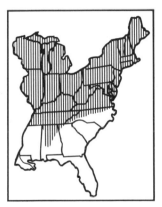

S. vitreum
(Mitchill)
Walleye

P. rex
Jordan and Evermann
Roanoke Perch

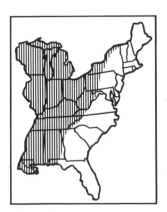

P. caprodes
(Rafinesque)
Logperch

P. shumardi
(Girard)
River Darter

    EE. Spinous dorsal without conspicuous blotches, or with basal blotches on all interradial membranes. Sides with about 9 conspicuous blotches and back with 4 dark saddles . . . . *P. uranidea*

  DD. Anal fin shorter, its posterior rays extending very slightly, if at all, behind posterior end of depressed soft dorsal

    E. Premaxillary frenum usually absent, or at least poorly developed . . . . *P. copelandi*

    EE. Premaxillary frenum present

      F. Caudal base with vertical row of 3 spots (lower two sometimes coalesced). Preopercle usually serrate. Gill membranes moderately cojoined

        G. Lateral line scales less than 65, tending toward vertical bars . . . . *P. nigrofasciata*

        GG. Lateral line scales more than 65

        H. Lateral line less than 75 scales . . . . *P. sciera*

        HH. Lateral line scales more than 75 . . . . *P. lenticula*

    FF. Caudal base with a single black spot or not spotted. Preopercle entire. Gill membranes usually separate

      G. Midventral scales on belly enlarged between pelvic fins only, not posteriad . . . . *P. aurantiaca*

      GG. Not as above

        H. Laterally 7 to 10 vertical bars which cross the back joining those of opposite side, or coalesce to form a lateral band

        I. 6 scales or less above lateral line and less than 50 in lateral line, 10-11 lateral bars which tend to coalesce, forming lateral band . . . . *P. crassa*

P. uranidea
(Jordan and Gilbert)
Stargazing Darter

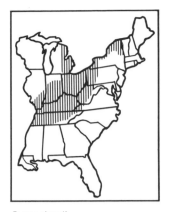

P. copelandi
(Jordan)
Channel Darter

P. nigrofasciata
(Agassiz)
Blackband Darter

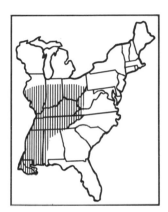

P. sciera
(Swain)
Dusky Darter

P. lenticula
Richards and Knapp
Freckled Darter

P. aurantiaca
(Cope)
Yellow Darter

P. crassa
(Jordan and Brayton)
Coarsescale Darter

II. 7 or more scales above, and more than 50 scales in lateral line. 7-10 distinct, well separated bars which are continuous over the back

    J. Subocular bar distinct and inclined downward and forward. . *P. evides*

    JJ. Subocular bar absent . . . . *P. palmaris*

HH. Pattern of round, horizontally oblong or vertical blotches essentially alternating with dorsal ones

    I. Lateral line scales usually less than 66. East of Appalachians

    J. Strong black median stripe on head behind chin . . . . . . *P. peltata*

    JJ. No such stripe . . . . *P. notogramma*

    II. Lateral line scales usually 66 or more. West of Appalachians

    J. Pattern of round, squarish or oblong blotches. Breast naked except usually a few enlarged scales between pelvics. Snout not notably long and pointed

      K. Lateral line scales 62-77. Cheeks and opercles scaly. No distinct black spot at soft dorsal origin . . . . *P. maculata*

      KK. Lateral line scales 88-90. Cheeks and opercles largely naked. Distinct black spot at soft dorsal origin. . . . . . . *P. macrocephala*

    JJ. Pattern of vertical elongate blotches at least anteriorly (except *P. squamata* which has a scaly breast). Snout long and pointed

P. evides
(Jordan and Copeland)
Gilt Darter

P. palmaris
(Bailey)
Bronze Darter

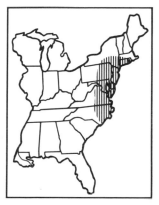

P. peltata
(Stauffer)
Shielded Darter

P. notogramma
(Raney and Hubbs)
Stripeback Darter

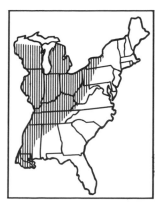

P. maculata
(Girard)
Blackside Topminnow

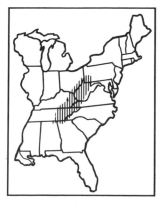

P. macrocephala
(Cope)
Longhead Darter

K. Breast with small scales in addition to large ones. Anal fin II, 9 or 10 . . . . *P. squamata*

KK. Breast naked except for enlarged scales, anal fin II, 8 or 9

L. Dorsal spines 11 or 12. Lateral line scales 59-79, scales above lateral line about 12. Eye length 1.1 to 1.5 in distance from union of gill membranes to tip of mandible . *P. phoxocephala*

LL. Dorsal spines usually 13 or 14. Lateral line scales 73 to 80, scales above lateral line 9 or 10. Eye length 1.5 to 1.8 in this distance . . . . *P. oxyrhyncha*

BB. Midline of belly with ordinary scales or naked only anteriad, premaxillary frenum present or absent

C. Body extremely long and slender, its depth contained more than seven times in standard length. Flesh translucent in life. Single anal spine

D. Premaxilla not protractile. Dorsal spines about 13 to 15. *Crystallaria asprella*

DD. Premaxilla protractile. Dorsal spines 7 to 11 . . . . *AMMOCRYPTA*

E. Body with at least 1 row of scales above and 2 rows below lateral line. Opercular spine present

F. Nape with at least a few scales near occiput. Opercle with a flattened, more triangular spine

G. Soft dorsal without dusky bar at base. Outer row of teeth in upper jaw scarcely enlarged . . . . *A. pellucida*

P. squamata
(Gilbert and Swain)
Olive Darter

P. phoxocephala
(Nelson)
Slenderhead Darter

P. oxyrhyncha
(Hubbs and Raney)
Longsnout Darter

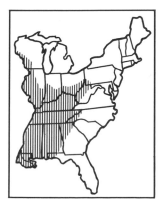

C. asprella
(Jordan)
Crystal Darter

A. pellucida
(Baird)
Eastern Sand Darter

   GG. Soft dorsal with a dusky bar at base. Outer row of teeth moderately enlarged . . . . *A. vivax*

  FF. Nape naked, or at most with a few scales on midline. Opercle with sharp, pin-like spine . . . . *A. clara*

 EE. Body naked anteriorly, except for lateral line scales and scales on caudal peduncle. Opercular spine obsolete . . . . *A. beani*

CC. Body not as long, depth contained fewer than seven times in standard length. Flesh opaque. 1 or 2 anal spines. . . . . *ETHEOSTOMA*

 D. Lateral line absent or with no more than 7 pored scales. Preoperculomandibular canal with 6 to 8 pores. Lateral line with no more than 37 scales. First dorsal with no more than 8 spines.

  E. Cheeks and opercles scaly. Lateral line with 3 to 7 pores . . . *E. proeliare*

  EE. Cheeks and opercles naked or nearly so. Lateral line absent or with only one or two pores . . . . *E. microperca*

 DD. Lateral line usually with 10 or more pores (if less the opercle is serrate) or otherwise not as above. Preoperculomandibular canal with 9 or more pores.

  E. Lateral line anteriorly curved upward, the least depth between lateral line and first dorsal base contained more than 4.2 times in distance below lateral line.

   F. Gill membranes completely separate. No blackish spots at tail base. 36 to 41 transverse scale rows . . . . *E. edwini*

  FF. Gill membranes at least slightly connected. Blackish spots present at tail base. Transverse scale rows 37 to 61

   G. Infraorbital canal complete

    H. Preopercle serrate. Infraorbital pores 6 . . . . *E. serriferum*

    HH. Preopercle entire. Pores 8 . . . . *E. gracile*

   GG. Infraorbital canal interrupted

A. vivax
Hay
Southwestern Sand Darter

A. clara
Jordan and Meek
Western Sand Darter

A. beani
Jordan
Naked Sand Darter

E. proeliare
(Hay)
Cypress Darter

E. microperca
Jordan and Gilbert
Least Darter

E. edwini
(Hubbs and Cannon)
Brown Darter

E. serriferum
(Hubbs and Cannon)
Sawcheek Darter

E. gracile
(Girard)
Slough Darter

H. Interorbital pores present . . . . *E. saludae*

HH. Interorbital pores absent

  I. Cheeks with embedded cycloid scales . . . . *E. collis*

  II. Cheek scales exposed, ctenoid

    J. Infraorbital pores 2 + 4 . . . . *E. zoniferum*

    JJ. Infraorbital pores 1 + 3, 1 + 4 or 2 + 3 . . . . *E. fusiforme*

EE. Lateral line nearly straight; this depth contained less than 4.2 times in distance below lateral line

F. Anus encircled by many fleshy villi. Body translucent in life . . *E. vitreum*

FF. Not as above

  G. Anal spines typically 1 (*E. nigrum* with characteristic "w" markings usually has 2 thin spines in Georgia and the Carolinas)

    H. Anal spine thin and flexible

      I. Lateral line nearly complete, extending beyond middle of soft dorsal. Dorsal fins close together (Evidence is accumulating for separation of eastern populations as *E. olmstedi*) . . . . *E. nigrum*

      II. Lateral line extending only to about middle of dorsal fins. Dorsal fins separated . . . . *E. chlorosomum*

    HH. Anal spine thick and heavy

      I. Lateral line complete, 3 middorsal saddles. Breast naked. *E. trisella*

      II. Lateral line incomplete. 4-6 middorsal saddles. Breast well scaled . . . . *E. tuscumbia*

  GG. Anal spines 2

    H. Both anal spines thin and flexible, the first usually less than 1/2 the length of the first ray

E. saludae
(Hubbs and Cannon)
Saluda Swamp Darter

E. collis
(Hubbs and Cannon)
Catawba Swamp Darter

E. zoniferum
(Hubbs and Cannon)
Banded Swamp Darter

E. fusiforme
(Girard)
Northern Swamp Darter

E. vitreum
(Cope)
Glassy Darter

E. nigrum
Rafinesque
Johnny Darter

E. chlorosomum
(Hay)
Bluntnose Darter

E. trisella
Bailey and Richards
Trispot Darter

E. tuscumbia
Gilbert and Swain
Spring Darter

I.   Pectoral fin longer than head. Anterior portion of belly naked or with only a few scales. Lateral line complete

  J.   Lateral line scales more than 50. Gill membranes scarcely connected, the distance from union of gill membranes to mandible tip contained 1.2 times in head length . . . .*E. perlongum*

  JJ. Lateral line scales less than 50. Gill membranes broadly connected; this distance contained 1.5 to 1.8 in head length

    K.   Snout pointed, with sloping profile. Saddles and blotches mostly indistinct . . . . *E. podostemone*

    KK. Snout profile nearly vertical. 7 lateral and 6 dorsal blotches distinct . . . . *E. longimanum*

II.   Pectoral fin shorter than head, anterior portion of belly fully scaled. Lateral line incomplete

  J.   Premaxilla protractile. Dorsal blotches broad at midline . . .
. . . . *E. stigmaeum*

  JJ. Premaxilla not protractile. Dorsal blotches hour-glass shaped, with narrowest portion on midline . . . . *E. jessiae*

HH. First anal spine much thicker and heavier than second, and usually more than 1/2 the length of the first ray

  I.   Premaxilla protractile

    J.   Pectoral fin length about equal to head length. 9 dorsal blotches, 9 or 10 vertical lateral blotches . . . . *E. atripinne*

    JJ. Pectoral fin longer than head. 7-8 dorsal blotches, lateral blotches, round or "w" shaped, or confluent

E. perlongum
(Hubbs and Raney)
Waccamaw Darter
(Waccamaw Lake)

E. podostemone
Jordan and Jenkins
Riverweed Darter

E. longimanum
Jordan
Longfin Darter

E. stigmaeum
(Jordan)
Speckled Darter

E. jessiae
(Jordan and Brayton)
Chickamauga Darter

E. atripinne
(Jordan)
Cumberland Snubnose Darter

K. Infraorbital dark bar curved downward and backward, lateral blotches round and distinct, but joined by narrow band. *E. simoterum*

KK. Infraorbital bar vertical, or inclined forward. Lateral blotches confluent, forming a lateral band, or "w" shaped . . . . . *E. duryi*

II. Premaxilla not protractile

J. Spiny dorsal with pronounced area of black pigment

K. Black blotch on posterior portion of dorsal . . . . . *E. luteovinctum*

KK. Black pigment on anterior portion of first dorsal fin

L. Black spot on first 2 interradial membranes. Nape scaly . . . . *E. thalassinum*

LL. Black spot on first 4-5 interradial membranes. Nape naked

M. Scales of back and sides without continuous stripes . *E. obeyense*

MM. Scales of back and sides with spots forming continuous . . . stripes . . . . *E. virgatum*

JJ. Spiny dorsal usually lacking pronounced black area

K. Two or three dark spots, or two spots fused into a bar at base of tail

E. simoterum
(Cope)
Tennessee Snubnose Darter

E. duryi
Henshall
Blackside Snubnose Darter

E. luteovinctum
Gilbert and Swain
Redbanded Darter

E. thalassinum
(Jordan and Brayton)
Seagreen Darter

E. obeyense
Kirsch
Banded Barcheek Darter

E. virgatum
(Jordan)
Streaked Barcheek Darter

    L.   Usually three dark spots at base of tail. Lateral line scales less than 70

      M.   Cheeks, opercles scaled. Usually 8-9 dorsal rays .*E. squamiceps*

     MM. Cheeks, opercles naked. 10-12 dorsal rays . . . . . *E. whipplei*

   LL. Two dark spots at base of tail, or spots fused into vertical bar. Lateral line scales more than 70. . . . . *E. sagitta*

KK. Caudal base usually not as above

    L.   Pectoral base black. Dorsal spines 10-12, rays 11-13. Dorsal blotches 6. Cheeks naked . . . . *E. rupestre*

   LL. Not with above combination of characters

    M.   No dorsal blotches or saddles

       N.   Opercle scaled. Dorsal spines 8, and ending in knobs . . . . *E. kennicotti*

      NN. Opercle naked. Dorsal spines 9 or more and not ending in knobs

        O.   Lateral line scales less than 43

         P.   Nape scaled. Breast with a few scales . . . . *E. mariae*

         PP. Nape with only a few scales. Breast naked. . . *E. fricksium*

        OO. Lateral line with 45 or more scales

         P.   No pronounced vertical bars on sides. With 12-13 dorsal rays. Lateral line complete . . . . *E. camurum*

E. squamiceps
Jordan
Spottail Darter

E. whipplei
(Girard)
Redfin Darter

E. sagitta
(Jordan and Swain)
Arrow Darter

E. rupestre
Gilbert and Swain
Rock Darter

E. kennicotti
(Putnam)
Stripetail Darter

E. mariae
(Fowler)
Cape Fear Darter

E. fricksium
Hildebrand
Savannah Darter

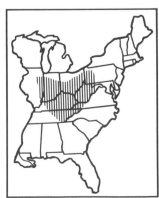

E. camurum
(Cope)
Bluebreast Darter

PP. Lacking this combination of characters

  Q.   Not more than 51 scales in lateral series

    R.   Lateral line complete. Sides with longitudinal stripes and 2-3 vertical blotches . . . . *E. rufilineatum*

    RR. Lateral line incomplete. Sides with small black spots . . . . *E. tippecanoe*

  QQ. At least 54 scales in lateral series

    R.   About 10 vertical bars on body . . . . *E. maculatum*

    RR. About 14 vertical bars on body . . . . *E. acuticeps*

MM. Dorsal blotches or saddles present (indistinct in *E. parvipinne*)

  N.   Breast scaled (*E. zonale,* which sometimes has the breast naked, keys out here. It has a combination of scaly cheeks and opercles, about 10 vertical bands laterally and about 6 dorsal saddles)

    O.   Space between pelvic fins very narrow . . . . . . *E. parvipinne*

    OO. Space between pelvic fins nearly as wide as base of fin . . . . *E. zonale*

E. rufilineatum
(Cope)
Redlined Darter

E. tippecanoe
Jordan and Evermann
Tippecanoe Darter

E. maculatum
Kirtland
Spotted Darter

E. acuticeps
Bailey
Sharphead Darter

E. parvipinne
Gilbert and Swain
Goldstripe Darter

E. zonale
(Cope)
Banded Darter

NN. Breast naked, or scaled only near pelvics

  O.   Lateral line incomplete

    P.   Dorsal blotches 6. Dorsal spines 8 and with knobs. . . . . *E. flabellare*

    PP. Dorsal blotches 7 or more. Dorsal spines lacking knobs

      Q.   Dorsal blotches 7-9

        R.   Lateral line with 55 to 66 scales . . . . *E. exile*

        RR. Lateral line with 40 to 47 scales . . . . *E. hopkinsi*

      QQ. Dorsal blotches about 10 (Most *E. asprigene* key out here. They have scaled cheeks and opercles.)

        R.   Pectoral rays 11-12. Gill membranes over-lapping and nearly separate. . . . . *E. spectabile*

        RR. Pectoral rays 13-14. Gill membranes broadly joined . . . . *E. caeruleum*

  OO. Lateral line complete or nearly so

    P.   Cheeks scaled

      Q.   Dorsal spines 12-14. Often with upper jaw concealed in groove under snout and "V" shaped markings laterally . . . . *E. blennioides*

      QQ. Dorsal spines 10-11

        R.   Belly naked. 8 anal rays . . . . *E. sellare*

        RR. Belly scaled. 6 or 7 anal rays

          S.   Lateral line with 47 to 55 scales . . . . . . *E. asprigene*

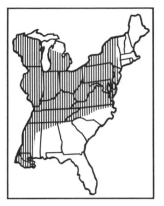

E. flabellare
Rafinesque
Fantail Darter

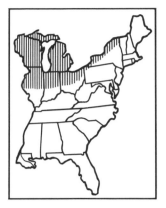

E. exile
(Girard)
Iowa Darter

E. hopkinsi
(Fowler)
Christmas Darter

E. spectabile
(Agassiz)
Orangethroat Darter

E. caeruleum
Storer
Rainbow Darter

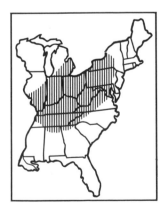

E. blennioides
Rafinesque
Greenside Darter

E. sellare
(Radcliffe and Welch)
Maryland Darter
(Swan Creek)

E. asprigene
(Forbes)
Mud Darter

        SS.   Lateral line scales less than 45

           T.   Preoperculomandibular pores 9. . . . . . *E. okaloosae*

          TT. Preoperculomandibular pores 10 . . . . . . *E. swaini*

PP. Cheeks naked

  Q.   Opercles scaled

    R.   Less than 56 lateral line scales

      S.   Belly scaled. Dorsal spines 10-12. . . . . . *E. jordani*

      SS. Belly naked. Dorsal spines 13 . . . . *E. histrio*

    RR. More than 56 lateral line scales

      S.   Four dorsal saddles. . . . . *E. cinereum*

      SS. 8-9 dorsal saddles . . . . *E. microlepidum*

  QQ. Opercles naked

E. okaloosae
(Fowler)

E. swaini
(Jordan)
Gulf Darter

E. jordani
Gilbert
Greenbreast Darter

E. histrio
Jordan and Gilbert
Harlequin Darter

E. cinereum
Storer
Ashy Darter

E. microlepidum
Raney and Zorach

R.   More than 62 scales in lateral line. . . . . . . *E. osburni*

RR. Less than 62 scales in lateral line

  S.   Combination of less than 49 lateral line scales and about 6 lateral blotches . . . . *E. inscriptum*

  SS. At least 49 lateral line scales *or* more than 7 lateral blotches or bars

    T.   Dorsal blotches 6

      U.   Scales present between pelvic fins . . . *E. kanawhae*

      UU. Breast completely naked. . . . .*E. swannanoa*

    TT. Dorsal blotches 4

      U.   Scales 42-45 . . . . *E. blennius*

      UU. Scales 50 to 58 . . . . *E. variatum*

E. osburni
(Hubbs and Trautman)
Finescale Saddled Darter

E. inscriptum
(Jordan and Brayton)
Altahama Darter

E. kanawhae
(Raney)
Kanawha Darter

E. swannanoa
Jordan and Evermann
Swannanoa Darter

E. blennius
Gilbert and Swain
Blenny Darter

E. variatum
Kirtland
Variegated Darter

STRUCTURES OF AMPHIBIANS

Some of the important characters that may give difficulty in keying of the amphibians are listed below.

Adpressed limbs. With the front legs pushed backwards and the back legs pushed forward along the sides of the body.

Bicornuate. With two lobes.

Canthus rostralis. Dorsolateral ridge of snout extending forward from eye to the nostril.

Costal grooves. Vertical grooves on the sides of many salamanders.

Dorsolateral folds. Folds of skin extending posteriorly from just behind the eye.

Intercalary cartilage. Extra element between the last and next-to-last segments of the toe of some frogs. Gives a stepped appearance to the toe.

Nasolabial groove. Tiny grooves extending from nasal openings to lips in salamanders of family Plethodontidae. Examine closely to see these.

Parotoid glands. Glandular swelling behind the eye.

Plica. A fold.

Reticulate. Having a netted pattern.

KEYS TO AMPHIBIANS

Frogs and toads are included in the key beginning on page 130, while the salamanders are included in the key beginning on page 140. The frogs and toads of the eastern United states lack tails, while the salamanders have them.

KEY TO FROGS AND TOADS

A. Transverse fold of skin across head behind eyes. Tympanum absent . . . . .
. . . . MICROHYLIDAE, *Gastrophryne carolinensis*

AA. No such fold. Tympanum present or absent

  B. Pupil of eye vertical. Pectoral glands on ventral surface near forelimbs. .
. . . . .PELOBATIDAE, *Scaphiopus holbrooki*

  BB. Pupil of eye rounded. No pectoral glands

    C. Tongue not bicornuate

      D. Parotoid glands present. Waist wide. Skin rough and with warts . . . .
. . . . BUFONIDAE, *BUFO*

        E. Parotoid glands as long as head . . . . *B. marinus*

        EE. Parotoid glands shorter than head

          F. Conspicuous light line from snout to vent. Adults under 35 mm . . . .
. . . . *B. quercicus*

          FF. No conspicuous light line from snout to vent or adults over 35 mm

            G. Parotoid glands small, subtriangular. Dark lateral stripe bordered
above by light . . . . *B. valliceps*

            GG. Parotoid glands large, not subtriangular; no dark lateral stripe bor-
dered above by light.

              H. Parietal cranial crest ending in a conspicuous knob. . . . . *B. terrestris*

              HH. Parietal crest not ending in a conspicuous knob

                I. Only one or two large warts in each of the largest of dorsal dark
spots. Often with numerous spots on breast and abdomen . . . .
. . . . *B. americanus*

                II. Three or more large warts in large dorsal dark spots. No spotting
or only with single "breast spot" . . . . *B. woodhousei*

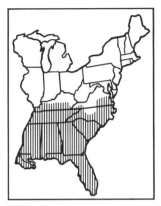

G. carolinensis
(Holbrook)
Narrow-mouthed Toad

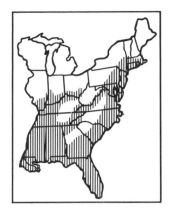

S. holbrooki
(Harlan)
Eastern Spadefoot

B. marinus
(Linnaeus)
Giant Toad

B. quercicus
Holbrook
Oak Toad

B. valliceps
Wiegmann
Gulf Coast Toad

B. terrestris
(Bonnaterre)
Southern Toad

B. americanus
Holbrook
American Toad

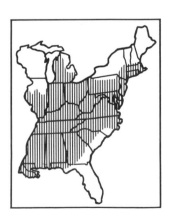

B. woodhousei
Girard
Woodhouse's Toad

DD. Parotoids absent. Waist narrow. Skin usually smooth

  E.   Intercalary bone absent . . . .
      . . . . LEPTODACTYLIDAE, *Eleutherodactylus planirostris*

  EE. Intercalary bone present, giving stepped appearance. . . . . HYLIDAE

    F.   Toe discs $\frac{1}{2}$ the diameter of tympanum or more . . . . *HYLA*

      G.   Black bordered light spot below posterior end of eye

        H.   Rear of thigh with green in life, skin on back smooth . . . . . *H. avivoca*

        HH. Rear of thigh with yellow or orange in life, skin of back smooth or
            with many small warts . . . . *H. versicolor*

     GG. No such spot

        H.   With X marking on back . . . . *H. crucifer*

        HH. Without X marking.

          I.   Finger discs approximately as large as tympanum. Thumb rudi-
             ment present. Rear of femur reticulate . . . . *H. septentrionalis*

          II. Discs distinctly smaller than tympanum. No thumb rudiment

            J.   Distinct broad dark or light stripe extending through or below eye
              and backward at least to the level of the groin. No distinct dorsal
              markings

             K.   Dark stripe bordered with white extending backward from
                 tympanum . . . . *H. andersoni*

E. planirostris
(Cope)
Greenhouse Frog

H. avivoca
Viosca
Bird-voiced Treefrog

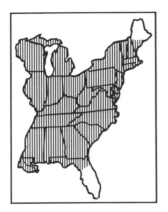

H. versicolor
Le Conte
Common Treefrog

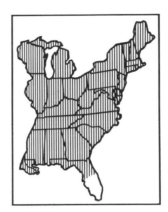

H. crucifer
Wied
Spring Peeper

H. septentrionalis
Boulenger
Cuban Treefrog

H. andersoni
Baird
Pine Barrens Treefrog

KK. Broad light stripe extending backward from just below eye . .
. . . . *H. cinerea*

JJ. Light stripe indistinct or not extending backward to the groin
when present or else dark dorsal markings present

K. Dorsum with many pronounced dark oval spots . . . . . *H. gratiosa*

KK. Markings on dorsum, if present, not as above

L. Concealed portion of thigh not dark with light spotting. *H. squirella*

LL. Concealed portion of thigh dark with small light spots *H. femoralis*

FF. Toe discs less than half the diameter of the tympanum

G. Hind feet well webbed. Posterior longitudinal femoral stripes present.
Usually warts on back . . . . *ACRIS*

H. First toe of hind leg partly free of webbing, 4th toe has 3 joints free
. . . . *A. gryllus*

HH. First toe completely webbed, except for disc, $1\frac{1}{2}$-2 joints of the
4th toe free . . . . *A. crepitans*

GG. Very little webbing. Stripes absent. Usually no warts

H. Body length not over 18 mm. Vomer without teeth . *Limnaoedus ocularis*

HH. Body length of adults greater than 18 mm. Vomer with teeth . . .
. . . .*PSEUDACRIS*

I. Dorsal pattern of 5 dorsal stripes, or 5 rows of spots (including
the eye stripes)

J. With the 3 median dorsal stripes lighter than the lateral ones. .
. . . . *P. brimleyi*

JJ. All five stripes of equal intensity

K. From below the Fall line; snout pointed; dorsal pattern of three
broad broken black stripes . . . . *P. nigrita*

H. cinerea
(Schneider)
Green Treefrog

H. gratiosa
Le Conte
Barking Treefrog

H. squirella
Latreille
Squirrel Treefrog

H. femoralis
Latreille
Piney Woods Treefrog

A. gryllus
(LeConte)
Southern Cricket Frog

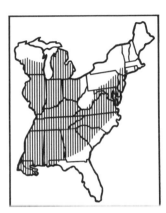

A. crepitans
Baird
Northern Cricket Frog

L. ocularis
(Holbrook)
Least Treefrog

P. brimleyi
Brandt and Walker
Brimley's Chorus Frog

P. nigrita
(LeConte)
Chorus Frog

KK. From above the Fall line or if from below, snout rounded; dorsal pattern of three narrow broken dark stripes (not black).
. . . . *P. triseriata*

II. Dorsal pattern not of five stripes or rows of spots

J. Back with spots or vermiculate or immaculate. . . . . . *P. streckeri*

JJ. Back with 4 dorsal stripes (including eye stripes)

K. With 2 broad dorsal stripes which approach or meet one another at middle of back. . . . . *P. brachyphona*

KK. Dorsal stripes, if well developed, not meeting on back, dark spots on sides and near groin . . . . *P. ornata*

CC. Tongue bicornuate behind. . . . . RANIDAE, *RANA*

D. Dark stripe through eye and tympanum (sometimes absent or obscure). .
. . . . *R. sylvatica*

DD. Without such stripes

E. With distinct dorsolateral folds

F. Dorsolateral folds extending about 2/3 length of body. No large dorsal spots . . . . *R. clamitans*

FF. Dorsolateral folds extending full length of body. Dorsal spotting usually well developed.

G. Rectangular spots in two rows between dorsolateral folds. Orange in groin and under legs in life . . . . *R. palustris*

GG. Spots not rectangular. No orange

H. Light colored line on upper jaw . . . . *R. pipiens*

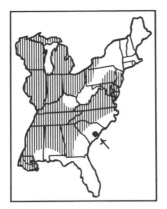

P. triseriata
(Wied)
Western Chorus Frog

P. streckeri
Wright and Wright
Strecker's Chorus Frog

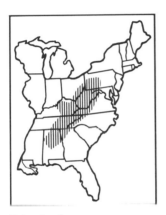

P. brachyphona
(Cope)
Mountain Chorus Frog

P. ornata
(Holbrook)
Ornate Chorus Frog

R. sylvatica
LeConte
Wood Frog

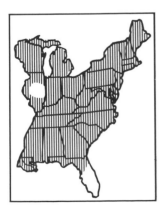

R. clamitans
Latreille
Green Frog

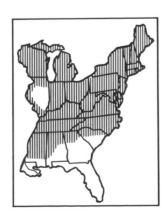

R. palustris
Le Conte
Pickerel Frog

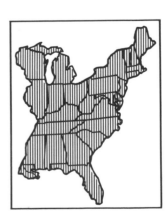

R. pipiens
Schreber
Leopard Frog

138

HH. No such line . . . . *R. areolata*

EE. Dorsolateral folds indistinct or absent

  F. Adults not over 75 mm. With reticulate dorsal pattern, or with dorso-lateral light stripes

    G. Usually a bright yellowish dorsolateral stripe extending backwards from eye. Venter spotted or reticulate . . . . *R. virgatipes*

    GG. No such stripes. Reticulate pattern dorsally. Venter immaculate
    . . . . *R. septentrionalis*

  FF. Adults over 75 mm. Rather uniform colored dorsally

    G. Rear of femur with horizontal pattern of light and dark blotches

      H. Fourth toe protrudes well beyond webbing. First finger generally less than second finger on front foot . . . . *R. grylio*

      HH. Fourth toe webbed to tip. First finger about equal to second finger in length. . . . . *R. catesbeiana*

    GG. Rear of femur without horizontal pattern of light and dark blotches
    . . . . *R. heckscheri*

R. areolata
Baird and Girard
Crawfish Frog

R. virgatipes
Cope
Carpenter Frog

R. septentrionalis
Baird
Mink Frog

R. grylio
Stejneger
Pig Frog

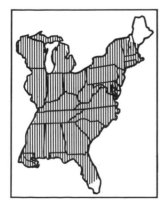

R. catesbeiana
Shaw
Bullfrog

R. heckscheri
Wright
River Frog

KEY TO SALAMANDERS

A.   Nasolabial grooves absent

  B.   Body eel-like with only one pair of limbs or two pairs of diminutive limbs

    C.   Anterior limbs only . . . . SIRENIDAE

      D.   3 pairs of gill slits, 4 toes . . . . *SIREN*

        E.   Costal grooves 36-39 . . . . *S. lacertina*

        EE. Costal grooves 31-36 . . . . *S. intermedia*

      DD. 1 pair of gill slits, 3 toes . . . . *Pseudobranchus striatus*

    CC. Two pairs of diminutive limbs . . . . AMPHIUMIDAE, *AMPHIUMA*

      D.   1 toe . . . . *A. pholeter*

      DD. 2 or 3 toes . . . . *A. means*

S. lacertina
Linnaeus
Greater Siren

S. intermedia
Le Conte
Lesser Siren

P. striatus
(Le Conte)
Dwarf Siren

A. pholeter
Neill

A. means
Garden
Amphiuma

BB. Body not eel-like

    C.   Adult animals with external gills. Toes 4-4 . . . . NECTURIDAE, *NECTURUS*

        D.   Dorsum with many black spots or dots. Body length to 300 mm

           E.   Adults seldom over 230 mm. Found on the North Carolina or gulf Coastal plains

              F.   Center of abdomen often unspotted. Gulf coast. . . . . *N. beyeri*

             FF. Belly completely spotted. Neuse and Tar River systems of North Carolina . . . . *N. lewisi*

          EE. Adults usually over 230 mm. Forms not of the area described above except along lower Mississippi River. . . . . *N. maculosus*

        DD. Few or no black spots or dots. Body length to 120 mm . . . . . . *N. punctatus*

  CC. No external gills

        D.   Size large, with wrinkled lateral skin fold . . . .
. . . . .CRYPTOBRANCHIDAE, *Cryptobranchus alleganiensis*

        DD. No lateral skin fold

           E.   Costal grooves indistinct . . . . . . .SALAMANDRIDAE, *NOTOPHTHALMUS*

N. beyeri
Viosca
Gulf Coast Waterdog

N. lewisi
Brimley
Neuse River Waterdog

N. maculosus
(Rafinesque)
Mudpuppy

N. punctatus
(Gibbes)
Dwarf Waterdog

C. alleganiensis
(Daudin)
Hellbender

F.   With dorsolateral stripe (red in life) not heavily bordered with black.
. . . . *N. perstriatus*

FF. Without dorsolateral stripe, or if present heavily bordered with black
. . . . *N. viridescens*

EE. Costal grooves distinct . . . . .AMBYSTOMATIDAE, *AMBYSTOMA*

F.   Plicae of tongue diverging from median furrow

  G.   Teeth on margin of jaw in single row . . . . *A. mabeei*

  GG. Teeth in more than one row

   H.   Strongly reticulate dorsal pattern . . . . *A. cingulatum*

   HH. Dorsum with gray lichen=like splotches or uniformly dark . . *A. texanum*

FF. Plicae not diverging from median furrow

  G.   Dorsum strongly marked with black and white . . . . *A. opacum*

  GG. Dorsum dark or with yellow or orange markings

   H.   Dorsum with yellow or orange

    I.   Rounded spots in dorsolateral series only . . . . *A. maculatum*

    II. Spots, bars, or blotches on back, sides, and belly . . . . *A. tigrinum*

   HH. Dorsum lacking yellow spots or blotches

    I.   10 costal grooves between fore and hind limbs. Body stout and
        short. . . . . *A. talpoideum*

    II. More than 10 costal grooves between fore and hind limbs (in some
        areas where the ranges of *A. jeffersonianum* and *A. laterale*
        approach each other or overlap, triploid females recognized as
        belonging to two different species are known. These are *A.
        tremblayi*, occurring with and resembling *A. laterale* and *A.
        platineum*, occurring with and resembling *A. jeffersonianum*.)

N. perstriatus
(Bishop)
Striped Newt

N. viridescens
(Rafinesque)
Newt

A. mabeei
Bishop
Mabee's Salamander

A. cingulatum
Cope
Flatwoods Salamander

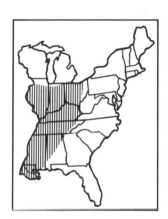

A. texanum
(Matthes)
Small-mouthed Salamander

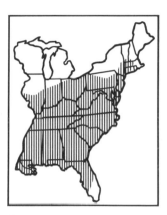

A. opacum
(Gravenhorst)
Marbled Salamander

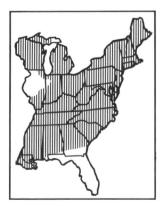

A. maculatum
(Shaw)
Spotted Salamander

A tigrinum
(Green)
Tiger Salamander

A. talpoideum
(Holbrook)
Mole Salamander

     J.   Mature adults 40-75 mm. in snout-vent length. Internarial distance 2.5 to 4.0 mm. Color black to gray with numerous large light flecks on sides. . . . . *A. laterale*

     JJ.   Mature adults 65-95 mm in snout-vent length. Internarial distance 4.5-6.5 mm. Color brownish-gray with a few small light flecks on sides . . . . *A. jeffersonianum*

AA. Nasolabial grooves present . . . . PLETHODONTIDAE

  B.   Gills present in adults

    C.   Pigment reduced or absent. Eyes vestigial, cave form . *Haideotriton wallacei*

    CC. Pigment present. Eyes developed . . . . *Gyrinophilus palleucus*

  BB. Gills absent in adults

    C.   Usually a light line from posterior corner of eye to angle of jaw or with internal nares concealed

      D.   Internal nares concealed . . . . *Leurognathus marmoratus*

      DD. Internal nares not concealed . . . . *DESMOGNATHUS* (This is a very difficult genus. There is much individual and age variation. Young individuals will not key.)

        E.   Tail oval or nearly circular in cross section

        F.   Snout-vent length under 32 mm in adults

          G.   Either a narrow middorsal line ending in "Y" to eyes or with indistinct dark smudges within the stripe . . . . *D. aeneus*

          GG. Not as above. Narrow herringbone pattern down back in light dorsal stripe. . . . . *D. wrighti*

        FF. Snout-vent length over 32 mm in adults. Pattern exceedingly variable, but often with line of distinct spots down back. . . . . *D. ochrophaeus*

A. laterale
Hallowell
Blue-spotted Salamander

A. jeffersonianum
(Green)
Jefferson's Salamander

H. wallacei
Carr
Georgia Blind Salamander

G. palleucus
McCrady
Tennessee Cave Salamander

L. marmorata
Moore
Shovel-nosed Salamander

D. aeneus
Brown and Bishop
Cherokee Salamander

D. wrighti
King
Pigmy Salamander

D. ochrophaeus
Cope
Mountain Salamander

EE. Tail sharp edged or keeled above

    F. Dorsum with strong alternate or opposite or fused markings. Venter pale. . . . . *D. monticola*

    FF. Not with above combination of characters

      G. Dorsum with strong alternate, opposite or fused markings, but venter mottled. Tail only weakly keeled . . . . *D. ocoee*

      GG. Dorsum with no distinctive pattern, or with pattern variable but usually not as above. If as above, then tail strongly keeled.

        H. Venter uniformly and deeply pigmented in large individuals; 2 rows of white spots along sides. Above the Fall line . . . *D. quadramaculatus*

        HH. Venter not as above, often mottled; if two rows of white spots, then from below the Fall line on the costal plain.

          I. 2-4 intercostal spaces between adpressed limbs. . . . . *D. fuscus*

          II. $4\frac{1}{2}$ to $5\frac{1}{2}$ intercostal spaces between adpressed limbs. . . *P. auriculatus*

CC. No light line from eye to angle of jaw. Internal nares open

  D. At least 90 mm snout-vent length in adults and with more than 9 costal folds between adpressed limbs . . . . *Phaeognathus hubrichti*

DD. Lacking this combination of characters

  E. Toes 4-4

    F. Tail with basal constriction. Belly white with black spots. . . . . *Hemidactylium scutatum*

    FF. No constriction. Belly not as above . . . . *Eurycea quadridigitatus*

  EE. Toes 5-4

    F. Tongue attached by central pedicel, free all around

      G. Vomerine and parasphenoid teeth forming continuous series

        H. Canthus rostralis marked by a light line. . . . *Gyrinophilus porphyriticus*

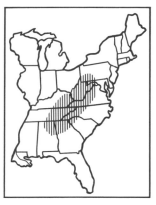

D. monticola
Dunn
Seal Salamander

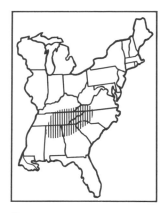

D. ocoee
Nicholls
Ocoee Salamander

D. quadramaculatus
(Holbrook)
Black-bellied Salamander

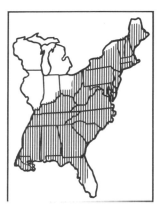

D. fuscus
(Rafinesque)
Dusky Salamander

D. auriculatus
(Holbrook)
Southern Dusky
    Salamander

P. hubrichti
Highton

H. scutatum
Schlegel
Four-toed Salamander

E. quadridigitatus
(Holbrook)
Dwarf Four-toed
    Salamander

G. porphyriticus
(Green)
Purple Salamander

HH. Canthus rostralis not marked by light line . . . . *PSEUDOTRITON*

    I.   Horizontal diameter of eye 1.2 to 1.5 in snout. Spots tending to remain separate . . . . *P. montanus*

    II.   Diameter of eye 1.5 to 2.0 in snout. Spots often fused . . . . *P. ruber*

GG. Vomerine and parasphenoid teeth not in continuous series. (Except in some *E. aquatica*) . . . . *EURYCEA* (part)

   H.   13 costal grooves . . . . *E. aquatica*

  HH. 14 or more costal grooves

    I.   Adpressed limbs separated by three or more costal folds. *E. bislineata*

    II.   Adpressed limbs separated by two or less costal folds

      J.   Black spots on orange ground color. No herringbone pattern ventrally on tail . . . . *E. lucifuga*

     JJ. Yellow ground color with middorsal dark stripe or with herringbone tail pattern or with much melanophore suffusion . . *E. longicauda*

FF. Tongue attached in front

   G.   Posterior edge of maxilla sharp-edged, without teeth. Phalanges square. . . . . *Aneides aeneus*

GG. Maxilla normal, with teeth. Phalanges not square

   H.   Conspicuous sensory pits on head; alternating light and dark longitudinal lines on sides . . . . *Stereochilus marginatus*

  HH. Without such pits or lines . . . . *PLETHODON*

    I.   5 or more costal folds between adpressed limbs

      J.   Reticulate pattern ventrally

        K.   Dorsal stripe broad, straight or finely serrate. No ventral orange pigment. Snout gray (some individuals lead-backed). . . . . *P. cinereus*

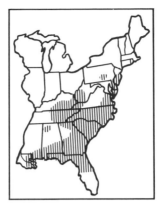

P. montanus
Baird
Mud Salamander

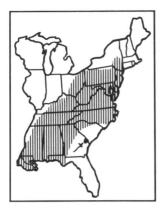

P. ruber
(Latreille)
Red Salamander

E. aquatica
Rose and Bush

E. bislineata
(Green)
Two-lined Salamander

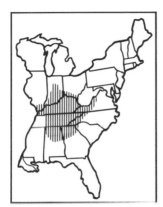

E. lucifuga
(Rafinesque)
Cave Salamander

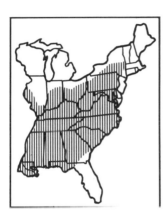

E. longicauda
(Green)
Long-tailed Salamander

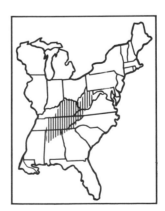

A. aeneus
(Cope)
Green Salamander

S. marginatus
(Hallowell)
Many-lined Salamander

P. cinereus
(Green)
Red-backed Salamander

KK. Dorsal stripe usually zig-zag. Ventral and humeral orange pigment. Orange snout . . . .*P. dorsalis*

JJ. Venter not reticulate . . . . *P. richmondi*

II. Less than 50 costal folds between adpressed limbs

J. Throat not conspicuously lighter than top of head. A black sala- mander usually spinkled dorsally and laterally with white spots . . . .*P. glutinosus*

JJ. Throat conspicuously lighter than top of head.

K. 1 costal fold or less between adpressed limbs. Chestnut dorsal stripe or spotting in life

L. Adpressed limbs usually not overlapping in adults. Chestnut stripe dorsally and white spots laterally but not dorsally. . . . . *P. yonahlossee*

LL. Adpressed limbs usually overlapping in adults. Chestnut spot- ting dorsally and white spots laterally and dorsally in life . . . . *P. longicrus*

KK. 1-4 costal folds between adpressed limbs. No chestnut stripe or spotting, but brassy spotting may be present

L. Size small, to 42 mm. Much dorsal brassy spotting; from NW N. C., Va., and NE Tenn. 3-4 costal folds between ad- pressed limbs . . . . *P. welleri*

LL. Not as above, larger. If much brassy spotting, then from NW South Carolina

M. With red legs or cheeks, or without conspicuous markings, or with brassy dorsum. If with white spots on sides from SW North Carolina . . . . *P. jordani*

MM. With small white or bluish spots usually restricted to lower side; if red pigment is present it occurs as small spots on the dorsum. Found in SW New York to Virginia. . . . *P. wehrlei*

P. dorsalis
Cope
Zigzag Salamander

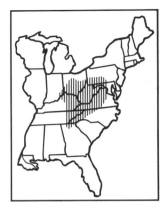

P. richmondi
Netting and Mittleman
Ravine Salamander

P. glutinosus
(Green)
Slimy Salamander

P. yonahlossee
Dunn
Yonahlossee Salamander

P. longicrus
Adler and Dennis

P. welleri
Walker
Weller's Salamander

P. jordani
Blatchley
Jordan's Salamander

P. wehrlei
Fowler and Dunn
Wehrle's Salamander

STRUCTURES OF REPTILES

Familiarity with Figures 4 and 5 and the terms listed will help in keying of the reptiles. The scales of a lizard are named essentially as are those of a snake, but more scales are often present.

Alveolar surface of jaw: The biting surface of turtle jaw.

Anal plate: The large single scale just ahead of the anus in snakes. It may be entire or diagonally divided.

Bridge (of turtles): Joint between the carapace and plastron.

Carapace: The upper portion of a turtle shell.

Frontonasal: Single large dorsal scale on head of lizard ahead of the frontal.

Plastron: The lower portion of a turtle shell.

Scale rows (counting): All scale rows are counted starting with one of the small scales next to the ventrals (in the middle of the body) and moving forward in a line over the dorsal portion of the body until the ventrals are reached on the other side.

Suboculars: Small scales below the eye and above the upper labials of lizards.

Symphysis of jaw: Meeting line of two halves of jaw.

Ventrals: The transversely elongate scales of the abdomen.

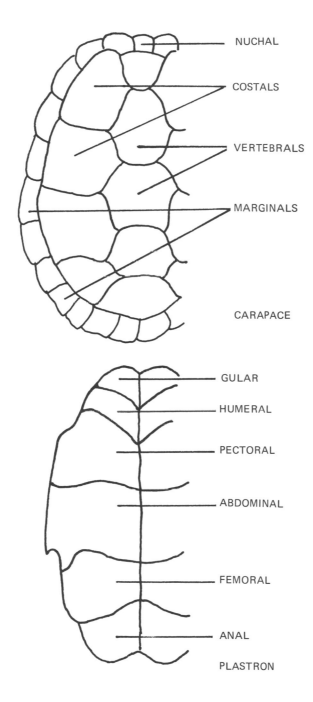

NUCHAL

COSTALS

VERTEBRALS

MARGINALS

CARAPACE

GULAR

HUMERAL

PECTORAL

ABDOMINAL

FEMORAL

ANAL

PLASTRON

Figure 4
STRUCTURE OF A TURTLE SHELL

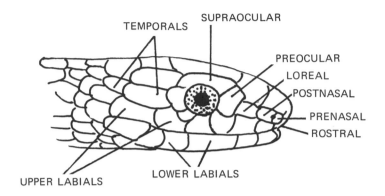

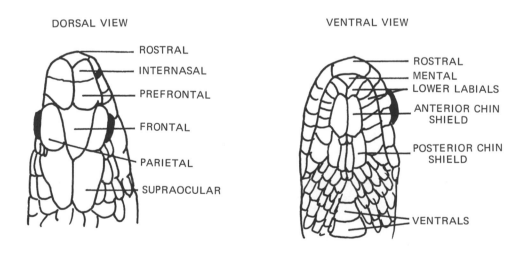

Figure 5
SCALES OF A SNAKE HEAD

KEYS TO REPTILES

Snakes can usually be separated from lizards by their lack of legs; most lizards have them. There are two genera of lizards in the eastern United States, *Ophisaurus* and *Rhineura,* which lack these appendages. *Ophisaurus* has a longitudinal fold of skin on each side of the body and has external ear openings. Snakes lack the fold and ear openings. *Rhineura* is earthworm like in appearance with the scales in rings around the body.

A.   Shell present . . . . Order CHELONIA (see key pg. 158)

AA. Shell lacking

  B.   Cloacal opening a longitudinal slit . . . . Order CROCODILIA

    C.   Snout broad and rounded, broader than interocular distance behind nostrils
    . . . .*Alligator mississipiensis*

    CC. Snout pointed, no wider than interocular distance for some distance behind
    nostrils . . . . *Crocodylus acutus*

  BB. Cloacal opening a transverse slit . . . . Order SQUAMATA

    C.   Legs and ear openings usually present (Legs absent in *Ophisaurus* which has
    a lateral longitudinal fold of skin; legs and ear openings absent in *Rhineura*
    which has the scales in earthworm like rings.) . . . .
    . . . . Suborder LACERTILIA (see key pg. 168)

    CC. Legs and ear openings absent . . . . . . .Suborder SERPENTES (see key pg. 174)

A. mississipiensis
(Daudin)
American Alligator

C. acutus
Cuvier
American Crocodile

KEY TO TURTLES, CHELONIA

A. Carapace covered with skin and with flexible edge. Claws 3-3 . . . .
. . . . TRIONYCHIDAE, *TRIONYX*

  B. Nasal septum with lateral ridge projecting into nostril

    C. Marginal ridge present . . . . *T. ferox*

    CC. Marginal ridge absent . . . . *T. spinifer*

  BB. Without such a ridge . . . . *T. muticus*

AA. Carapace covered with shields. Claws not 3-3

  B. Plastron with 11 or fewer shields. Pectoral shield not forming part of bridge

    C. Tail long with a crest of horny tubercles. Rear of carapace strongly serrate
. . . . .CHELYDRIDAE

      D. Single row of marginal shields. Tail with two rows of large scales be-
neath. . . . . *CHELYDRA*

        E. Back of head with flat plates. Dorsal surface of neck with rounded tuber-
cles. . . . . *C. serpentina*

        EE. Back of head with granular scales and scattered tubercles. Neck with
long pointed tubercles . . . . *C. osceola*

      DD. 5th to 8th marginal shields doubled. Tail with numerous small scales be-
neath. . . . . *Macroclemys temmincki*

    CC. Tail short, without crest. Rear of carapace nearly smooth. Plastron with
11 plates . . . . .KINOSTERNIDAE

      D. Anterior and posterior edges of pectoral shields at an angle to one another,
forming a "K". Plastron hinged . . . .*KINOSTERNON*

        E. Carapace with 3 longitudinal yellow lines. . . . . *K. bauri*

        EE. Without such lines

          F. Ninth marginal plate much higher than eighth. . . . .*K. flavescens*

          FF. Ninth marginal plate about equal to eighth in height. . . . . . .*K. subrubrum*

      DD. Edges of pectoral shields nearly parallel. Not forming a "K". No hinges
. . . . *STERNOTHAERUS*

T. ferox
Schneider
Florida Softshell

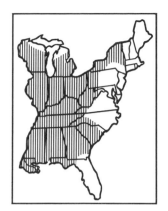

T. spinifer
LeSueur
Spiny Softshell

T. muticus
(LeSueur)
Smooth Softshell

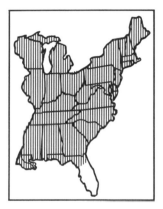

C. serpentina
Linnaeus
Snapping Turtle

C. osceola
Stejneger
Florida Snapping Turtle

M. temmincki
(Troost)
Alligator Snapping Turtle

K. bauri
Garman
Striped Mud Turtle

K. flavescens
(Agassiz)
Yellow Mud Turtle

K. subrubrum
(Lacepède)
Mud Turtle

E. Barbels on throat and chin. Side of head often with light stripes. Shields not overlapping . . . . *S. odoratus*

EE. Barbels on chin only. Light stripes alternating with dark stripes when present. Shields overlapping

    F. Gular absent. Carapace with sharp median keel which slopes sharply to marginals . . . . *S. carinatus*

    FF. Gular present. Carapace not sloping to marginals. If keel is present, then not sloping sharply

        G. Head with network of dark lines. Carapace low and flattened. No sharp middorsal keel. . . . . *S. depressus*

        GG. Head with dark stripes or spots. Sharp middorsal keel or 3 keels. Shell high and rounded . . . . *S. minor*

BB. Plastron with 12 shields, pectoral shield forming part of bridge

C. Top of head with scales . . . . TESTUDINIDAE, *Gopherus polyphemus*

CC. Top of head with skin only . . . . EMYDIDAE

  D. Plastron with transverse hinge or hinges

    E. Front of upper jaw curved downward as viewed from side . . . . *TERRAPENE*

      F. Second vertebral keeled. Interfemoral suture less than half the length of the interabdominal suture. . . . . *T. carolina*

      FF. No keel. Interfemoral suture more than half the length of the interabdominal suture. . . . . *T. ornata*

    EE. Front of upper jaw curved upward. . . . . *Emydoidea blandingi*

  DD. Plastron without hinges

    E. Axillaries and inguinals rudimentary or absent

      F. Front of upper jaw, as viewed from side, curved downward . . . *CLEMMYS*

        G. Carapace with yellow spots. No median keel . . . . *C. guttata*

        GG. No such spots. Median keel usually present

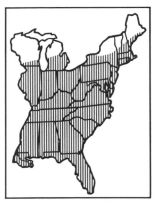

S. odoratus
(Latreille)
Stinkpot

S. carinatus
(Gray)
Keel-backed Musk Turtle

S. depressus
Tinkle and Webb
Flattened Musk Turtle

S. minor
(Agassiz)
Loggerhead Musk Turtle

G. polyphemus
(Daudin)
Gopher Tortoise

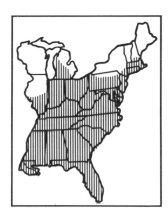

T. carolina
(Linnaeus)
Box Turtle

T. ornata
(Agassiz)
Western Box Turtle

E. blandingi
(Holbrook)
Blanding's Turtle

C. guttata
(Schneider)
Spotted Turtle

      H.   Orange blotch on temple. No notches between posterior marginals. Ridges on carapace plates indistinct . . . . *C. muhlenbergi*

    HH. No blotch. Notches between posterior marginals. Prominent ridges on carapace plates . . . . *C. insculpta*

  FF. Front of jaw curved upward . . . . *Malaclemys terrapin*

EE. Axillaries and inguinals well developed

  F.   Each posterior marginal with deep notch; if not then length of symphysis equal to or greater than $\frac{1}{2}$ the length of intergular suture . . *GRAPTEMYS*

    G.   Lower jaw symphysis longer than least distance between orbits. Triangular yellow spot behind eye . . . . *G. geographica*

    GG. Symphysis equalling or shorter than this distance, or if longer, then triangular spot lacking

      H.   Central light blotch on each costal shield . . . . *G. flavimaculata*

      HH. Without such blotches

        I.   Costal shields with complete light circles. Postocular yellow mark not more than half the eye diameter . . . . *G. oculifera*

        II.   Not as above

          J.   Yellow or greenish blotch behind eye which is larger than area of orbit

            K.   Irregular bordered yellow bar across ventral surface of lower jaw . . . . *G. barbouri*

           KK. No such bar, although an elongate blotch may be present. *G. pulchra*

C. muhlenbergi
(Schoepff)
Bog Turtle

C. insculpta
(Le Conte)
Wood Turtle

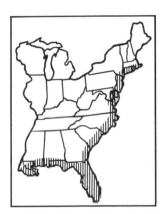

M. terrapin
(Schoepff)
Diamondback Terrapin

G. geographica
(Le Sueur)
Map Turtle

G. flavimaculata
Cagle
Yellow-blotched Sawback

G. oculifera
Baur
Ringed Sawback

G. barbouri
Carr and Marchand
Barbour's Map Turtle

G. pulchra
Baur
Alabama Map Turtle

        JJ.   No such blotch

           K.   Postorbital line extending anteriorly under the eye. . . . . . .*G. kohni*

           KK. No such line

              L.   Vertical posteriorly curved yellow line behind the eye which joins a diagonal line on upper surface of head. Spines of vertebral shield broadened, knoblike . . . . *G. nigrinoda*

              LL. Oval, rectangular or comma-shaped spot behind eye. Spines of vertebral shields not knoblike. . . . . . . *G. pseudogeographica*

FF. No such notches

   G.   Upper jaw with notch bordered by cusps

     H.   Ridge of alveolar surface of upper jaw without serrations . . . . *Chrysemys picta*

   HH. With serrations

     I.   No markings on bridge or only a few large spots or bars . . . . *Pseudemys nelsoni*

     II.   Markings on bridge oblong or in a concentric pattern with light centers. . . . . *Pseudemys rubriventris*

 GG. Upper jaw smooth or if with notch, notch not bordered by cusps

     H.   First marginal not extending beyond suture between first costal and first vertebral . . . . *Deirochelys reticularia*

   HH. First marginal extending beyond this suture. . . . . . .*PSEUDEMYS* (part)

G. kohni
Baur
Mississippi Map Turtle

G. nigrinoda
Cagle
Black-knobbed Sawback

G. pseudogeographica
Gray
False Map Turtle

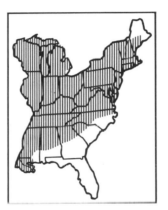

C. picta
(Schneider)
Painted Turtle

P. nelsoni
Carr
Florida Red-bellied Turtle

P. rubriventris
(Le Conte)
Red-bellied Turtle

D. reticularia
(Latreille)
Chicken turtle

I. Ridge of alveolar surface of upper jaw without serrations but often with tubercles. Often a spot or line behind eye which is at least half the eye diameter . . . . *P. scripta*

II. Ridge serrated. Spots or lines less than half the eye diameter if present

    J. Light "C" shaped figure on second costal scute. Plastron with numerous dark markings . . . . *P. concinna*

    JJ. Light vertical line or lines on second costal scute. Plastron unmarked or only lightly patterned . . . . *P. floridana*

P. scripta
(Schoepff)
Pond Slider

P. concinna
(Le Conte)
River Cooter

P. floridana
(Le Conte)
Cooter

KEY TO LIZARDS, LACERTILIA

A. Legs absent

   B. Scales in earthworm like rings around body. Ear opening absent. Longitudinal fold absent . . . . .AMPHISBAENIDAE, *Rhineura floridana*

   BB. No such rings. Ear opening present. Longitudinal fold extending the length of the body . . . . .ANGUIDAE, *OPHISAURUS*

      C. 1 or 2 upper labials on each side in contact with orbit. Frontonasal usually double . . . . *O. compressus*

      CC. Upper labials separated from orbit by scales. Frontonasal single

         D. White markings on posterior corners of scales. No distinct middorsal stripe present. No dark stripes on belly . . . . *O. ventralis*

         DD. White markings in middle of scales, often forming stripes. A distinct middorsal stripe usually present in adults, always in young. Dark stripes usually present on belly. . . . . *O. attenuatus*

AA. Legs present

   B. Eyelids absent, pupil vertical, head covered with small, granular scales

      C. Digits either not widened, or, if widened, with a single round scale at tips of toes. . . . . SPHAERODACTYLIDAE

         D. Digits not widened . . . . *Gonatodes fuscus*

         DD. Digits widened at tips with a single round scale on each toe tip . . . .*SPHAERODACTYLUS*

           E. Dorsal scales rather large and strongly keeled, overlapping and larger than ventral scales . . . . *S. notatus*

           EE. Dorsal scales smaller than ventral scales, granular and not overlapping . . . . *S. cinereus*

      CC. Digits widened with a series of flattened lamellae, and with last joint appearing clawlike . . . . .GEKKONIDAE, *Hemidactylus turcicus*

R. floridana
Baird
Worm Lizard

O. compressus
Cope
Island Glass Lizard

O. ventralis
(Linnaeus)
Eastern Glass Lizard

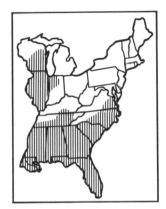

O. attenuatus
Baird
Slender Glass Lizard

G. fuscus
(Hallowell)
Yellow-headed Gecko

S. notatus
Baird
Reef Gecko

S. cinereus
Wagler
Ashy Gecko

H. turcicus
(Linnaeus)
Mediterranean Gecko

BB. Eyelids present, movable; pupil not vertical, head often with larger and more variable scales

   C.   All scales around body about equal in size. Scales smooth, shiny . . SCINCIDAE

     D.   Legs very short with not more than 2 digits . . . . *Neoseps reynoldsi*

    DD. Legs longer, each with five digits

      E.   No supranasals (scale between nasal and single frontonasal). Translucent disc on lower eyelid . . . . *Scincella laterale*

     EE. Supranasals present. No such disc, scaly eyelids . . . . *EUMECES*

       F.   Postnasal absent

         G.   Dorsolateral light stripes extending through second scale row from middorsal line. No median light line on head or body. . . . . . .*E. egregius*

        GG. Dorsolateral light stripes absent or not including second scale row at a point above forelimb. Median light stripe present or absent. . . .
          . . . . *E. anthracinus*

      FF. Postnasal present

        G.   Subcaudals not or only scarcely widened transversely . . *E. inexpectatus*

        GG. Subcaudals transversely widened

         H.   No postlabials, or 1 or 2 of very small size. Usually 8 upper labials, 4th ahead of rather than below the eye . . . . *E. laticeps*

         HH. 2 postlabials. Usually 7 upper labials, 4th below the eye. . *E. fasciatus*

CC. Scales around body not about equal in size

N. reynoldsi
Stejneger
Sand Skink

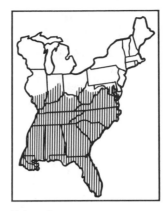

S. laterale
(Say)
Ground Skink

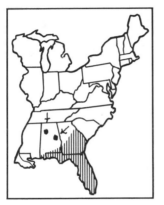

E. egregius
(Baird)
Florida Red-tailed Skink

E. anthracinus
(Baird)
Coal Skink

E. inexpectatus
Taylor
Southeastern Five-lined Skink

E. laticeps
Schneider
Broad-headed Skink

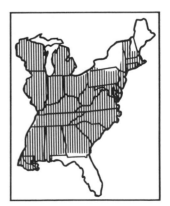

E. fasciatus
(Linnaeus)
Five-lined Skink

D. Belly scales smaller than, or less than 3 times larger than back scales .
. . . . IGUANIDAE

E. Scales on next to last phalanx of digits widened. . . . . *ANOLIS*

F. Pattern of four poorly defined chevrons pointing toward the rear. Line across top of head between eyes. . . . . *A. distichus*

FF. No chevrons. No line

G. Tail with dorsal ridge. Ventral scales much larger than dorsal scales. . . . . *A. sagrei*

GG. Tail without a dorsal ridge. Ventral scales equal to or somewhat larger than dorsal scales. . . . . *A. carolinensis*

EE. These scales not widened

F. Single, central, dorsal, longitudinal row of large keeled scales. Nasal in contact with rostral . . . . *Leiocephalus carinatus*

FF. No single, central, dorsal, longitudinal row of enlarged scales . . . .
. . . . *SCELOPORUS*

G. Lateral dark stripe distinct. Dorsal pattern indistinct. . . . . . . *S. woodi*

GG. No distinct lateral dark stripe or 2 dark lateral stripes . . . *S. undulatus*

DD. Belly scales more than five times larger than back scales . . . .
. . . . TEIDAE, *Cnemidophorus sexlineatus*

A. sagrei
Cocreau
Brown Anole

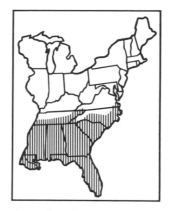

A. carolinensis
Voigt
Green Anole

A. distichus
Cope
Bahaman Bark Anole

L. carinatus
Gray
Bahama Curl-tailed Lizard

S. woodi
Stejneger
Florida Scrub Lizard

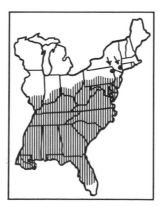

S. undulatus
Latreille
Eastern Fence Lizard

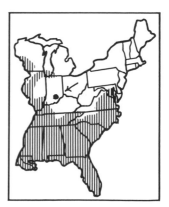

C. sexlineatus
(Linnaeus)
Six-lined Racerunner

KEY TO SNAKES, SERPENTES

A.   Pit present between eye and nasal opening . . . . .CROTALIDAE

  B.   Rattle present

    C.   Top of head covered with large scales . . . .*SISTRURUS*

      D.   Upper preocular not in contact with postnasal . . . . *S. miliaris*

      DD. Upper preocular in contact with postnasal . . . . *S. catenatus*

    CC. Top of head with small scales . . . . *CROTALUS*

      D.   Pattern of distinct diamond-shaped marks with light centers and yellow
          borders. Vertical line present on posterior edge of prenasals and first
          upper labials. . . . . *C. adamanteus*

      DD. Pattern of chevron shaped crossbands, or pattern absent. No vertical
          light line . . . .*C. horridus*

  BB. Rattle absent . . . . *AGKISTRODON*

    C.   Loreal scale present. 23 scale rows . . . . *A. contortrix*

    CC. Loreal scale absent. 25 scale rows . . . . *A. piscivorus*

AA. Pit absent

  B.   A pair of permanently erect grooved fangs in front of upper jaw. Black
    yellow and red rings around body with red and yellow rings touching.   Nose
    black . . . . ELAPIDAE, *Micrurus fulvius*

  BB. No such fangs. Without pattern described above . . . . .COLUBRIDAE

    C.   Anal plate not divided

      D.   Scales keeled

        E.   27 or more scale rows. More than 2 prefrontals . . *Pituophis melanoleucus*

        EE. Less than 27 scale rows. 2 prefrontals

          F.   Two rows of triangular black spots down middle of belly. . . . .
          . . . . *Tropidoclonion lineatum*

S. miliaris
(Linnaeus)
Pigmy Rattlesnake

S. catenatus
(Rafinesque)
Massasauga

C. adamanteus
Beauvois
Eastern Diamondback
    Rattlesnake

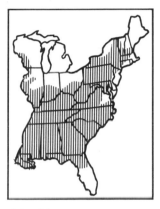

C. horridus
Linnaeus
Timber Rattlesnake

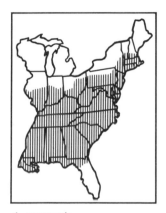

A. contortrix
(Linnaeus)
Copperhead

A. piscivorus
(Lacepede)
Cottonmouth

M. fulvius
(Linnaeus)
Coral Snake

P. melanoleucas
(Daudin)
Pine Snake

T. lineatum
(Hallowell)
Lined Snake

FF. Lacking such spots . . . . *THAMNOPHIS*

    G.   lateral stripe involving 4th scale row above belly anteriorly

        H.   Tail usually more than 27% of total length

            I.   Parietal spots large and fused across interparietal suture . . . .
. . . . *T. proximus*

           II. Parietal spots, if present, small and clearly separated by inter-
parietal suture . . . . *T. sauritus*

      HH. Tail usually less than 27% of total length

           I.   Lateral stripe anteriorly on scale rows 3 and 4 . . . . *T. radix*

           II. On rows 2, 3, and 4

              J.   Upper labials usually 6. Scale rows usually 17 in middle of
body . . . . *T. brachystoma*

             JJ. Upper labials usually 7. Scale rows usually 19 . . . . . . *T. butleri*

      GG. Lateral stripe involving scale rows 2 and 3, or sometimes absent. .
. . . . *T. sirtalis*

DD. Scales smooth

    E.   Loreal absent . . . . *Stilosoma extenuatum*

    EE. Loreal present

        F.   Rostral large, projecting beyond lower jaw. Ventral surface light
without markings. . . . . *Cemophora coccinea*

        FF. Rostral smaller. Ventral surface with at least some dark markings

           G.   17 scale rows. . . . . *Drymarchon corais*

          GG. More than 17 scale rows . . . . *LAMPROPELTIS*

T. sauritus
(Linnaeus)
Ribbon Snake

T. proximus
(Say)
Western Ribbon Snake

T. radix
(Baird and Girard)
Plains Garter Snake

T. brachystoma
(Cope)
Short-headed Garter Snake

T. butleri
(Cope)
Butler's Garter Snake

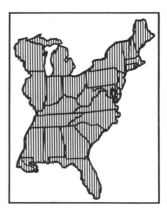

T. sirtalis
(Linnaeus)
Common Garter Snake

S. extenuatum
Brown
Short-tailed Snake

C. coccinea
(Blumenbach)
Scarlet Snake

Drymarchon corais
(Daudin)
Indigo Snake

    H.   Pattern without red, and without rings or black bordered blotches . . . . *L. getulus*

   HH. Pattern with red, or with rings or black bordered blotches

      I.   Pattern of rings, or if of blotches, then these broadly in contact with 5th or lower row of scales. . . . . *L. triangulum*

     II.   Pattern of black edged dorsal blotches extending no lower than upper edge of 5th row of scales. . . . . *L. calligaster*

CC. Anal plate divided

  D.   Either the preocular or the loreal missing; only one scale between the nasal scale and the eye

    E.   Preocular present and not horizontally elongate.   Loreal absent

     F.   Scales smooth . . . . *TANTILLA*

       G.   Light neck band crossing tips of parietals. . . . . *T. coronata*

      GG. Not crossing tips of parietals . . . . *T. gracilis*

    FF. Scales keeled. . . . . *STORERIA*

     G.   Scales in 15 rows. *S. occipitomaculata*

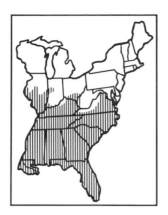

L. getulus
Linnaeus
Common Kingsnake

L. triangulum
Lacepede
Milk Snake

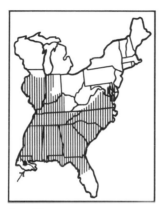

L. calligaster
(Harlan)
Prairie Kingsnake

T. coronata
Baird and Girard
Crowned Snake

T. gracilis
Baird and Girard
Flat-headed Snake

S. occipitomaculata
(Storer)
Red-bellied Snake

      GG. Scales in 17 rows . . . . *S. dekayi*

  EE. Loreal present and horizontally elongate. Preocular absent

    F.   13 scale rows . . . . *Carphophis amoenus*

    FF. More than 13 scale rows

      G.   Lower labials not more than 7, scale rows 15 or 17 . . . . *VIRGINIA*

        H.   Scales strongly keeled. 1 postocular. 5 upper labials . . . *V. striatula*

        HH. Scales smooth or nearly so. 2 postoculars. 6 upper labials. *V. valeriae*

      GG. Lower labials 8 to 10. Scale rows generally 19 at midbody . . *FARANCIA*

        H.   Single internasal . . . . *F. abacura*

        HH. 2 internasals . . . . *F. erytrogrammus*

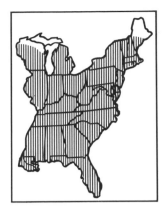

S. dekayi
(Holbrook)
Brown Snake

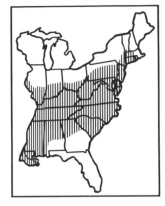

C. amoenus
(Say)
Worm Snake

V. striatula
Linnaeus
Rough Earth Snake

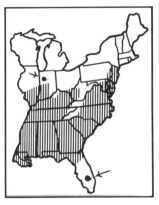

V. valeriae
(Baird and Girard)
Smooth Earth Snake

F. abacura
(Holbrook)
Mud Snake

F. erytrogrammus
(Palisot de Beauvois)
Rainbow Snake

DD. Loreal and preocular present, thus with two scales between nasal scale and eye

    E.   Rostral keeled . . . . *HETERODON*

        F.   Prefrontals touching one another . . . . *H. platyrhinos*

        FF. Prefrontals separated by small scales

          G.   Belly immaculate or mottled with gray-brown. . . . . *H. simus*

          GG. Belly black with small yellow patches. . . . . *H. nasicus*

    EE. Rostral not keeled

        F.   Scales, at least the dorsal ones, strongly keeled

          G.   Scale rows 17, bright green color in life (usually bluish gray when preserved) . . . . *Opheodrys aestivus*

          GG. Scale rows more than 17

           H.   Scale rows 19

             I.   Lower labials 7. 1 preocular . . . . *Clonophis kirtlandi*

             II.   Lower labials 9-11.   Usually 2 preoculars . . . . *REGINA* (part)

               J.   1 long dark median stripe on belly, or no markings except at ends of the ventrals . . . . *R. grahami*

               JJ. 2 long dark median stripes near middle of belly, at least anteriorly

                 K.   Light stripes present at sides of belly . . . . *R. septemvittata*

                 KK. No light stripes present at sides of belly . . . . *R. rigida*

         HH. Scale rows more than 19 . . . . *NATRIX*

           I.   Scale rows usually 27 to 33 (If 25, a pattern of alternating dorsal and lateral spots present.).   Lower labials usually 11 to 13

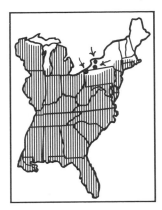

H. platyrhinos
Latreille
Eastern Hognose Snake

H. simus
(Linnaeus)
Southern Hognose Snake

H. nasicus
Baird and Girard
Western Hognose Snake

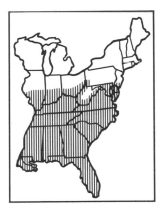

O. aestivus
(Linnaeus)
Rough Green Snake

C. kirtlandi
(Kennicott)
Kirtland's Water Snake

R. grahami
(Baird and Girard)
Graham's Water Snake

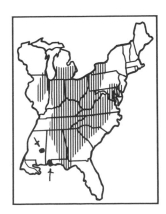

R. septemvittata
(Say)
Queen Snake

R. rigida
(Say)
Glossy Water Snake

    J.   One or more suboculars. *N. cyclopion*

    JJ. No suboculars, eye in contact with upper labials

       K.   2 or more anterior temporals. Rear portion of parietals broken into small scales. . . . . *N. taxispilota*

       KK. 1 anterior temporal. Rear portion of parietals not as above. . . . . *N. rhombifera*

  II. Scale rows usually 21 to 25. Lower labials usually 10

    J.   Ventral surface with conspicuous, dark, crescentic or vermiculate markings, or with row of light spots

       K.   Dark crossbands anteriorly, but alternating with lateral blotches posteriorly. Head usually unicolored or irregular. . . . *N. sipedon*

       KK. Dark crossbands throughout length of body; no alternation with lateral blotches. Dark stripe from eye to angle of jaw. . *N. fasciata*

    JJ. Ventral surface without markings, or with dark anterolateral margins of ventrals . . . . *N. erythrogaster*

FF. Scales smooth or dorsal scales may be weakly keeled

  G.   Lower preocular small and wedged between two of the upper labials

    H.   Tail very long and thin. Eye over upper labials 4 and 5. Generally 8 upper labials . . . . *Masticophis flagellum*

    HH. Tail not as long, more normal in appearance. Eye over upper labials 3 and 4, generally 7 upper labials . . . . . . *Coluber constrictor*

  GG. Lower preocular of normal size and position

N. cyclopion
(Dumeril, Bibron and Dumeril)
Green Water Snake

N. taxispilota
(Holbrook)
Brown Water Snake

N. rhombifera
Hallowell
Diamond-backed Water Snake

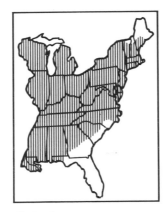

N. sipedon
(Linnaeus)
Common Water Snake

N. fasciata
(Linnaeus)
Banded Water Snake

N. erythrogaster
(Forster)
Plain-bellied Water Snake

M. flagellum
(Shaw)
Coachwhip

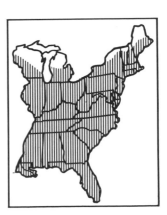

C. constrictor
Linnaeus
Racer

H.   More than 23 scale rows . . . . *ELAPHE*

  I.   Neck bands crossing parietals and uniting on frontals . . . . *E. guttata*

 II.   Not as above

  J.   More than 220 ventrals . . . .*E. obsoleta*

 JJ.   Less than 220 ventrals . . . . *E. vulpina*

HH. Less than 23 scale rows

  I.   Single internasal. At least 5 rows of scales in anal region with
       keels. . . . .*Regina alleni*

 II.   Two internasals. Scales usually all smooth (except in some
       Seminatrix, which also may have keeled scales in anal region).

  J.   2 preoculars, neck ring or black spots on belly or both . . . . .
       . . . . *Diadophis punctatus*

 JJ.   Usually 1 preocular, not patterned as above

   K.   15 scale rows. Nasal plate undivided. . . . . . *Opheodrys vernalis*

  KK. 17 scale rows. Nasal plate divided

   L.   Dark line from rostral through eye to last upper labial. 7
        upper labials. More than 60 caudals . . . . . . *Rhadinea flavilata*

  LL. No such line. Usually 8 upper labials. Less than 60 caudals
       . . . . *Seminatrix pygaea*

E. guttata
(Linnaeus)
Corn Snake

E. obsoleta
(Say)
Rat Snake

E. vulpina
(Baird and Girard)
Fox Snake

R. alleni
(Garman)
Striped Swamp Snake

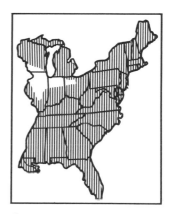

D. punctatus
(Linnaeus)
Eastern Ringneck Snake

O. vernalis
(Harlan)
Smooth Green Snake

R. flavilata
(Cope)
Yellow Lipped Snake

S. pygaea
Cope
Black Swamp Snake

STRUCTURES OF MAMMALS

The classification of mammals is based in great part on dentition and bones of the skull, hence one must become thoroughly acquainted with the general characters of teeth and skulls to identify mammals.

Certain standard measurements are made of mammals before they are skinned. These are: total length (from tip of nose to end of tail vertebrae, not to end of hairs), tail length (from base to end of vertebrae), and hind foot length (from ankle to end of longest toenail). These are then placed on the specimen label. If measurements on a label read 150, 35, 20, it is understood that these are total length, tail length, and hind foot length. It is meaningless to measure prepared skins. Use the original measurements taken by the collector.

The following definitions and figures will be helpful in using the keys to identify mammals.

Calcar. A small bone protruding inward from the hind leg along the edge of the intermoral membrane in bats.

Canal. An elongate tunnel-like opening through thick bone (see foramen).

Emarginate. Notched.

Foramen. An opening through a thin plate of bone.

Forearm. The main supporting bone of the wing in bats.

Interfemoral membrane. The membrane connecting the hind legs and tail in bats.

Lophs. Transversely elongate cusps of teeth.

Molariform teeth. Premolars are distinguished from molars by being preceded by milk teeth. Molars are not. This cannot be determined by examination of the adult skull, hence premolars and molars are considered together as molariform teeth and include all teeth behind the canines.

Patagium. A flap of skin, specifically the one connecting the front and hind legs in flying squirrels.

Postglenoid length of skull. The distance from the posterior end of the glenoid fossa (where the lower jaw articulates with the skull) to the posterior end of the skull.

Rooted teeth. Teeth with closed roots, or "rooted teeth" do not continue to grow, but mature relatively early in the life of the mammal. These usually can be recognized by the lack of parallel striae, and because the sides of the teeth are not parallel. They contract towards the base. The presence of cusps indicates rooted teeth, but some rooted teeth have loops and triangles rather than characteristic cusps (Fig. 7).

Rootless teeth. Rootless teeth generally continue to grow throughout the life of the animal. Parallel striae can be seen laterally. Rootless teeth have loops and triangles but not cusps (Fig. 7).

Sagittal crest. Median ridge on the braincase.

Tragus. Prounounced lobe protruding upwards inside a bat ear.

Unicuspids. In shrews, the group of small teeth between the large bilobed first incisor, and the group of large molariform teeth.

Zygomatic breadth. Greatest width across the zygomatic arches.

190

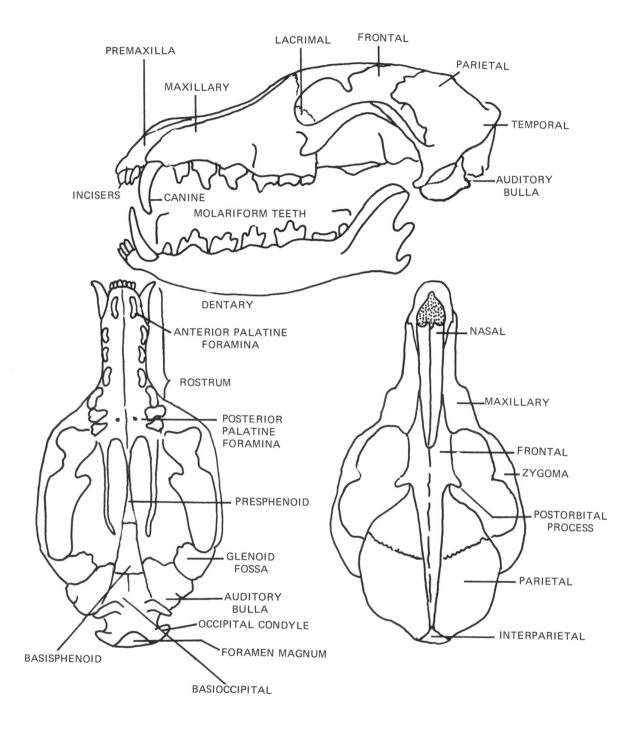

PREMAXILLA
LACRIMAL
FRONTAL
PARIETAL
MAXILLARY
TEMPORAL
AUDITORY
BULLA
INCISERS
CANINE
MOLARIFORM TEETH
DENTARY
ANTERIOR PALATINE
FORAMINA
ROSTRUM
POSTERIOR
PALATINE
FORAMINA
PRESPHENOID
GLENOID
FOSSA
AUDITORY
BULLA
OCCIPITAL CONDYLE
BASISPHENOID
FORAMEN MAGNUM
BASIOCCIPITAL
NASAL
MAXILLARY
FRONTAL
ZYGOMA
POSTORBITAL
PROCESS
PARIETAL
INTERPARIETAL

Figure 6
SKULL OF A MAMMAL

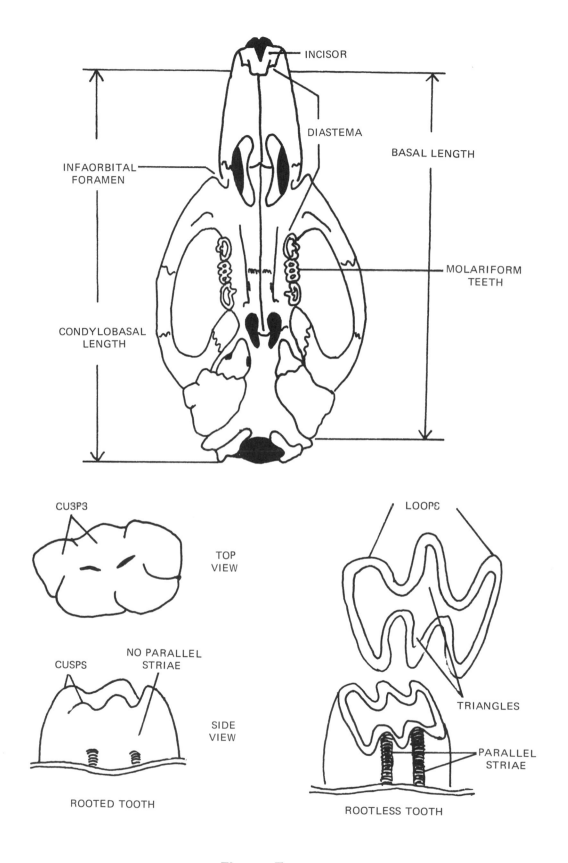

Figure 7

KEYS TO MAMMALS

A. Inner toe of hind foot without nail and opposable. Five upper incisors on each side. . . . . MARSUPIALIA, DIDELPHIDAE, *Didelphis marsupialis*

AA. Inner toe of hind foot, if present, with nail and not opposable. Fewer than 5 upper incisors on each side.

  B. Body covered with bony plates forming a hard shell dorsally. No teeth present in anterior one-fourth of upper or lower jaw . . . .
. . . . EDENTATA, DASYPODIDAE, *Dasypus novemcinctus*

  BB. Body covered with fur. At least incisors present in anterior one-fourth of lower jaw.

    C. Fingers longer than forearm and supporting a membrane which serves as a wing. Incisors of the separate sides of the upper jaw separated by an emargination of the palate (except in *Eumops*). . CHIROPTERA (see key pg. 206)

    CC. Fingers shorter than forearm and not supporting a flight membrane. Incisors of the two halves of upper jaw, if present, in single line

      D. Animals with hooves. Skull with upper canines and incisors lacking . . .
. . . . ARTIODACTYLA, CERVIDAE

      E. Antlers palmate . . . . *Alces alces*

      EE. Antlers not palmate.

        F. Posterior narial cavity divided by vomer . . . . *Odocoileus virginianus*

        FF. Posterior narial cavity not completely divided by vomer. *Cervus canadensis*

      DD. Animals with claws. Diastema may be present, but incisors present

        E. Diastema present, the canine teeth absent

          F. Two upper incisors. Ears shorter than combined length of tail vertebrae . . . . RODENTIA (see key pg. 212)

          FF. Four upper incisors. Ears longer than this length. . . . .
. . . . LAGOMORPHA, LEPORIDAE

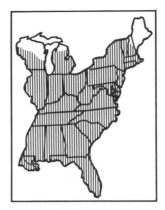

D. marsupialis
Linnaeus
Common Opossum

D. novemcinctus
Linnaeus
Nine-banded Armadillo

A. alces
(Linnaeus)
Moose

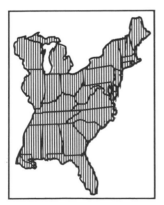

O. virginianus
(Zimmermann)
Whitetail Deer

Cervus canadensis
Erxleben
Wapiti

194

G. Interparietals fused with parietals. Hind foot more than 110 mm . *LEPUS*

  H. Pelage brownish or grayish

    I. Tail dark all around. . . . . *Lepus americanus*

    II. Tail black above, light below, or all white

     J. Tail black above, white below. . . . . *L. europaeus*

    JJ. Tail all white . . . . *L. townsendi*

  HH. Pelage all white

    I. Ear from notch more than 82 dry (87 fresh), least interorbital breadth more than 26 . . . . *L. townsendi*

    II. Ear from notch less than 82 dry and 87 fresh, least interorbital breadth less than 26. *L. americanus*

GG. Interparietals distinct. Hind foot less than 110 mm . . . . . *SYLVILAGUS*

  H. Anterior extension of supraorbital process present. Posterior portion free of braincase or separated by a slit from braincase. Nape rich cinnamon. . . . . *S. floridanus*

  HH. Anterior extension of supraorbital process absent (or if a point is present, then at least 5/6 of posterior extension is fused to braincase)

    I. Basilar length of skull more than 63 mm. . . . . *S. aquaticus*

    II. Basilar length of skull less than 63 mm

     J. Underside of tail white. Posterior extension of supraorbital process tapering to a slender point, this point free of braincase or barely touching it and leaving a slit . . . . *S. transitionalis*

    JJ. Underside of tail brown or gray. Posterior extension of supraorbital process fused to skull, usually for entire length . .*S. palustris*

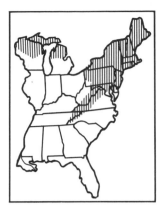

L. americanus
Erxleben
White-tailed Jackrabbit

L. europaeus
Pallas
European Hare

L. townsendi
Bachman
White-tailed Jackrabbit

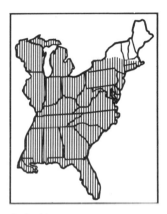

S. floridanus
(Allen)
Eastern Cottontail

S. aquaticus
(Bachman)
Swamp Rabbit

S. transitionalis
(Bangs)
Allegheny Cottontail

S. palustris
(Bachman)
Marsh Rabbit

EE. Diastema absent, canines and incisors present

F. Total length more than 200 mm. Canine teeth well developed as "eye" teeth, longer than premolars or incisors . . .CARNIVORA (see key pg. 200)

FF. Total length less than 200 mm. Canine teeth not usually much longer than premolars or incisors . . . . INSECTIVORA

G. Forefeet more than twice as wide as hind feet. Eyes hidden. . TALPIDAE

H. Tail more than 60 mm. Snout with 22 fleshy tentacles. Third incisor resembling a canine. Premaxillaries extending well forward of the narial aperture. Unicuspids separated . . . . *Condylura cristata*

HH. Tail shorter. No tentacles. Third incisor not resembling a canine

I. 10 teeth above on each side. Complete auditory bullae. First upper incisor simple. Tail nearly naked. . . . . . . *Scalopus aquaticus*

II. 11 teeth above on each side. Incomplete auditory bullae. First upper incisor with accessory cusp. Tail hairy . . *Parascalops breweri*

GG. Forefeet less than twice as wide as hind feet. Eyes small but visible. Teeth with chestnut tips. Zygoma absent . . . . SORICIDAE

H. Tail short, less than 30 mm. Either four unicuspids present, or the lateral edge of the braincase produced into a sharp angle

I. Color grayish. Total length more than 90 mm. 5 unicuspids present, the fifth hidden behind a projection of the fourth (the form found in Dismal swamp, Virginia, is currently recognized as a separate species, *B. telmalestes*) . . . . *Blarina brevicauda*

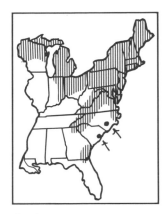

C. cristata
(Linnaeus)
Star-nosed Mole

S. aquaticus
(Linnaeus)
Common Mole

P. breweri
(Bachman)
Hairy-tailed Mole

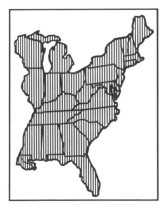

B. brevicauda
(Say)
Short-tailed Shrew

II. Color brownish. Total length less than 90 mm. Four unicuspids present, the fourth hidden . . . . *Cryptotis parva*

HH. Tail long, more than 30 mm. Five unicuspids present

I. Unicuspids 3 and 5 reduced in size. Animal very small, weighing 2-5 grams. Very difficult to distinguish from *Sorex cinereus* except by dentition. . . . . *Microsorex hoyi*

II. Only unicuspid 5 notably reduced in size. Animals vary in size from 2.5 to 20 or more grams . . . . SOREX

J. Color generally grayish (old *Sorex fumeus* may be brownish)

K. Total length greater than 140 mm . . . .*S. palustris*

KK. Total length less than 140 mm

L. Infraorbital foramen with posterior border lying behind space between first and second upper molars. Tail more than 50 mm . . . . *S. dispar*

LL. Infraorbital foramen with posterior border lying ahead of the space. Tail less than 50 mm . . . . *S. fumeus*

JJ. Color brown

K. Distinctly tricolored, with back dark, sides lighter, and belly still lighter. . . . .*S. arcticus*

KK. Without tricolor pattern; brown above and light below

L. Rostrum long and narrow. Greatest width across outside of first large molariform teeth usually more than 2.0 in distance from posterior end of palate to anterior end of first incisors. Third upper unicuspid not smaller than fourth. Inner ridge of upper unicuspids with pigment . . . .*S. cinereus*

LL. Rostrum short, wider. This width usually less than 2.0. Third upper unicuspid usually smaller than fourth. Inner ridge of upper unicuspid lacking pigment . . . . . . *S. longirostris*

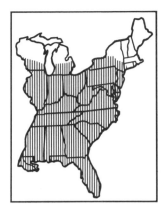

C. parva
(Say)
Least Shrew

M. hoyi
(Baird)
Pigmy Shrew

S. palustris
Richardson
Water Shrew

S. dispar
Batchelder
Gray Shrew

S. fumeus
Miller
Smoky Shrew

S. arcticus
Kerr
Arctic Shrew

S. cinereus
Kerr
Masked Shrew

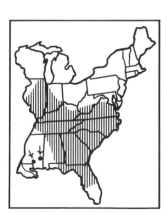

S. longirostris
Bachman
Long-nosed Shrew

KEY TO CARNIVORA

A.   Feet modified into flippers. Hind limbs directed backwards and used for swimming. Incisors 3/2 .PHOCIDAE, *Phoca vitulina*

AA. No flippers, hind limbs normal. Incisors 3/3

  B.   6 upper molariform teeth. Seven lower molariform teeth. Either very large with vestigal tail, or doglike

    C.   Skull with long narrow rostrum. Upper tooth rows not parallel. Tail long and bushy . . . . CANIDAE

      D.   Postorbital processes thickened, convex dorsally. Basal length of skull more than 147 mm . . . . *CANIS* (including *C. familiaris*)

        E.   Greatest length of skull usually more then 250 mm . . . . *C. lupus*

        EE. Greatest length of skull usually less than 250 mm . . . . *C. latrans*

      DD. Postorbital processes thin, concave dorsally. Basilar length of skull usually less than 147 mm

        E.   Sagittal crest U shaped. Tail not white tipped . . . *Urocyon cinereoargenteus*

        EE. Sagittal crest V shaped. Tail white tipped . . . . *Vulpes vulpes*

    CC. Skull with broad rostrum. Upper tooth rows parallel. Animal very large and with vestigial tail . . . . URSIDAE, *Ursus americanus*

  BB. Not as above. Not doglike. Tail well developed

    C.   6 upper and 6 lower molariform teeth. Black facial mask. Large bushy and ringed tail . . . . PROCYONIDAE, *Procyon lotor*

P. vitulina
Linnaeus
Harbor Seal

C. lupus
Linnaeus
Gray Wolf

C. latrans
Say
Coyote

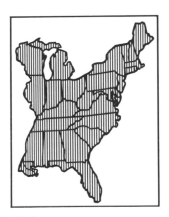

U. cinereoargenteus
(Schreber)
Gray Fox

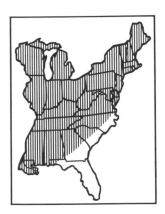

V. vulpes
Linnaeus
Red Fox

U. americanus
Pallas
Black Bear

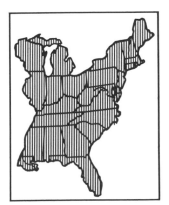

P. lotor
(Linnaeus)
Raccoon

CC. Not as above

    D.   Molariform teeth either 4/3 or 3/3. Form catlike; with very short tail,
           or else animal very large (body length 1075 to 1375 mm) . . . . FELIDAE

      E.   4 upper molariform teeth. Tail long . . . *Felis concolor* (and *F. domesticus*)

     EE. 3 upper molariform teeth. Tail short . . . . *LYNX*

       F.   Tip of tail black on top and bottom . . . . *L. canadensis*

      FF. Tip of tail black only on top . . . . *L. rufus*

DD. More than 3 lower molariforms. 4 or 5 upper molariforms . . .MUSTELIDAE

      E.   Animals black and white. Palate not extending appreciably beyond
           posterior edge of last upper molars. Molariform teeth 4/5

       F.   Four or more lines of broken stripes or spots. Total length less than
           500 mm. Top of skull flat. . . . . *Spilogale putorius*

       FF. Back with continuous white stripes. Total length more than 500 mm.
           Top of skull as seen in profile with an angle of about 60°. .*Mephitis mephitis*

     EE. Animals not black and white. Palate extends appreciably beyond
         posterior edge of last upper molars

       F.   Feet broad and webbed. Molariform teeth 5/5, skull flat and broad. .
          . . . . *Lutra canadensis*

       FF. Feet not broad and webbed. Teeth not 5/5

        G.   Molariform teeth 5/6. Hind foot at least 75 mm long. Top of head
           brown . . . . *MARTES*

F. concolor
Linnaeus
Mountain Lion

L. canadensis
Kerr
Lynx

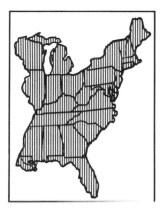

L. rufus
(Schreber)
Bobcat

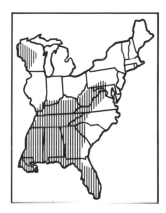

S. putorius
(Linnaeus)
Eastern Spotted Skunk

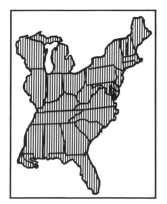

M. mephitis
(Schreber)
Common Striped Skunk

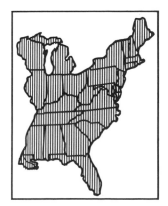

L. canadensis
(Schreber)
River Otter

H. Orange on throat and chest. Skull rounded behind. Greatest length of skull less than 95 mm . . . . *M. americana*

HH. No orange. Skull angular behind and more than 95 mm . . . *M. pennanti*

GG. Molariform teeth 4/5. Hind foot less than 75 mm or else a white stripe on head

H. Braincase triangular. Skull more than 90 mm long. Last molar triangular. White stripe on head. Hind foot more than 75 mm. . .
. . . . *Taxidea taxus*

HH. Braincase elongate. Skull less than 90 mm long. Last molar dumbell shaped. No white stripe on head. Hind foot less than 75 mm .
. . . . *MUSTELA*

I. Tail not black tipped

J. Size small. Less than 300 mm total length. Skull less than 40 mm long. Tail about one inch in length . . . . *Mustela nivalis*

JJ. Size larger. More than 300 mm total length. Skull more than 40 mm long. Tail much more than one inch . . . . . . *Mustela vison*

II. Tail black tipped

J. Postglenoid length of skull less than 47% of condylobasal length.
. . . . *M. frenata*

JJ. Postglenoid length of skull more than 47% of condylobasal length
. . . . *M. erminea*

M. americana
(Turton)
Marten

M. pennanti
(Erxleben)
Fisher

T. taxus
(Schreber)
Badger

M. nivalis
Linnaeus
Least Weasel

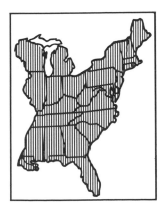

M. vison
Schreber
Mink

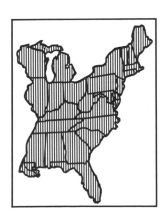

M. frenata
Lichtenstein
Long-tailed Weasel

M. erminea
Linnaeus
Short-tailed Weasel

KEY TO CHIROPTERA

A. One incisor on each side of upper jaw. Tail extending beyond the interfemoral membrane, *or* interfemoral membrane furred wholly or on basal half (if it is furred on the basal half and the bat is dark brown with silver-tipped dorsal hairs, take AA), *or* bat dull brown and hair sparse (*Nycticeius,* difficult to separate from *Myotis,* except by skull)

  B. Upper surface of interfemoral membrane wholly furred, 5 upper molariform teeth, but the first reduced in size and inside the canine. Skull very short, squarish . . . . VESPERTILIONIDAE (in part) *LASIURUS*

    C. Forearm more than 45 mm. Dorsal hairs mostly tipped with white . . . . *Lasiurus cinereus*

    CC. Forearm less than 45 mm. Color mahogany brown, red, orange, or yellowish

      D. Color red, orange, or yellowish . . . . *L. borealis*

      DD. Mahogany brown . . . . *L. seminola*

  BB. Upper surface of interfemoral membrane not furred or furred only on basal half. All molariform teeth in normal tooth row

    C. Tail extending far beyond end of interfemoral membrane. Five upper molariform teeth . . . . MOLOSSIDAE

      D. Bony palate with emargination back of incisors. Forearm not over 58 mm . . . . *Tadarida brasiliensis*

      DD. No emargination. Forearm over 58 mm . . . . *Eumops glaucinus*

    CC. Tail not extending far beyond interfemoral membrane. 4 upper molariform teeth . . . . VESPERTILIONIDAE (in part)

      D. Upper surface of interfemoral membrane furred on basal half. Color light yellowish brown. Skull short, depth of braincase including bullae, approximately half greatest length . . . . *Dasypterus floridanus*

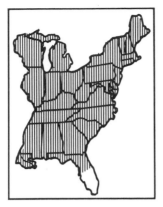

L. cinereus
(Beauvois)
Hoary Bat

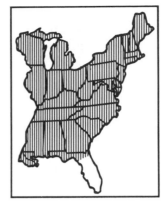

L. borealis
(Miller)
Red Bat

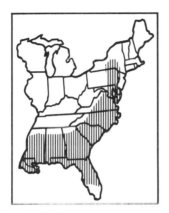

L. seminola
(Rhoads)
Seminole Bat

T. brasiliensis
(I. Geof. St.-Hilaire)
Freetail Bat

E. glaucinus
(Wagner)
Eastern Mastiff Bat

D. floridanus
Miller
Florida Yellow Bat

DD. Membrane not furred. Color dull brown. Fur short, sparse. Skull not so deep. . . . . *Nycticeius humeralis*

AA. Two upper incisors on each side. Tail not extending noticeably beyond inter-femoral membrane *or* membrane not furred or if furred on basal half, color very dark. . . . . VESPERTILIONIDAE (in part)

  B. 4 upper molariform teeth on each side, all about the same size. Forearm at least 43 mm long. Ears less than 30 mm long. Color light brown. . . .
  . . . . *Eptesicus fuscus*

BB. 5 or 6 upper molariform teeth. Forearm less than 43 mm long or else with ears at least 30 mm long

  C. 5 upper molariform teeth. First one reduced in size. Ears very large, or fur distinctly tricolor, or interfemoral membrane furred above on basal half

    D. Skull large, about 15 mm in length, dorsal aspect of skull very convex or "humped," auditory bullae very large. Ears at least 30 mm in length
    . . . . *PLECOTUS*

      E. Tips of ventral hairs buff in color. No accessory cusp on first incisor
      . . . . *Plecotus townsendi*

      EE. Tips of ventral hairs white. Often an accessory cusp on first incisor .
      . . . . *P. rafinesquii*

    DD. Skull not humped or else skull less than 14 mm long

      E. Skull very flat or concave dorsally as viewed from side. Dorsal fur very dark, with silver tips . . . . *Lasionycteris noctivagans*

      EE. Skull very small, less than 14 mm in length and convex in lateral view. Fur generally reddish, and distinctly tricolored . . . . *Pipistrellus subflavus*

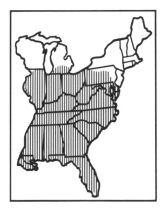

N. humeralis
Rafinesque
Evening Bat

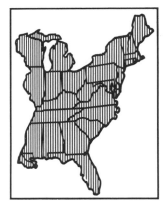

E. fuscus
(Beauvois)
Big Brown Bat

P. townsendi
Cooper
Townsend's Big-eared Bat

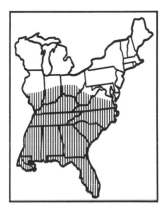

P. rafinesquii
Lesson
Rafinesque's Big-eared Bat

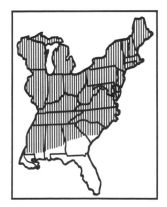

L. noctivagans
(Le Conte)
Silver-haired Bat

P. subflavus
(Cuvier)
Eastern Pipistrelle

CC. Six upper molariform teeth, first two reduced in size. Bats generally
small and brownish in color . . . . *MYOTIS* (identification of species is difficult)

D. Wing insertion is at the tarsus instead of at side of foot (normally seen only
in fresh or alcoholic specimens). Dorsal fur is uniform in color throughout, rather than being dark at base. . . . . *M. grisescens*

DD. Wing insertion is at the side of the foot. Dorsal fur dark at base.

E. Very small *Myotis* (forearm 30-36 mm, hind foot generally less than
7 mm). Light in color, but with black lips, and black mask. Keeled
calcar . . . . *M. subulatus*

EE. Larger; no black mask

F. Pinkish brown coloration. Keeled calcar. In hibernation is found in
very tight clusters . . . . *M. sodalis*

FF. Color generally lighter

G. Ears extending noticeably beyond nostrils when laid forward. Tragus
sickle-shaped. . . . . *M. keenii*

GG. Ears extending to nostrils when laid forward. Tragus straight

H. Dorsal fur dense, wooly. Sagital crest present in adults. . . . . . .
. . . . *M. austroriparius*

HH. Dorsal fur normal, silky, no sagital crest. . . . . *M. lucifugus*

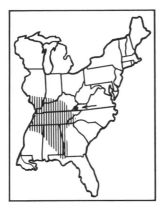

M. grisescens
Howell
Gray Bat

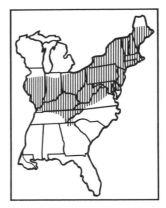

M. subulatus
(Say)
Lieb's Bat

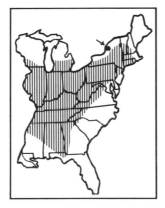

M. sodalis
Miller and Allen
Social Bat

M. keenii
(Merriam)
Keen's Bat

M. austroriparius
(Rhoads)
Southeastern Bat

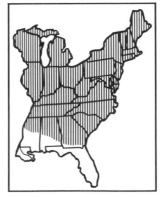

M. lucifugus
(Le Conte)
Little Brown Bat

KEY TO RODENTIA

A. Infraorbital opening larger than foramen magnum. Large rodent with stiff spine-like bristles . . . . ERETHIZONTIDAE, *Erethizon dorsatum*

AA. Infraorbital opening smaller than foramen magnum. No stiff spine-like bristles

  B. Infraorbital opening minute, not accomodating major muscles. Tail either well furred or greatly flattened. If not, then animal has external, fur lined cheek pouches.

    C. Tail well furred. No external cheek pouches. Skull with prominent postorbital processes. . . . . SCIURIDAE

      D. Incisors white. Top of skull flat. Supraorbital processes at right angles to skull. Tail short (less than $\frac{1}{4}$ of total length). Animals large, averaging 6-10 lbs . . . . *Marmota monax*

     DD. Incisors yellow. Top of skull convex. Supraorbital process not at right angle to skull. Tail more than $\frac{1}{4}$ of total length. Animals smaller, weighing less than 4 lbs.

       E. "V" shaped notch in skull directly over orbits. Lateral furred patagium present. . . . . *GLAUCOMYS*

         F. Brownish pelage. Total length of animal generally over 260 mm. Greatest length of skull generally over 36 mm . . . . *G. sabrinus*

        FF. Grayish pelage. Total length generally less than 260 mm. Skull less than 36 mm (The species of *Glaucomys* are difficult to distinguish) *G. volans*

      EE. No such notch. No patagium

       F. Infraorbital opening a foramen rather than a canal. Either two or four light stripes running lengthwise on the back

E. dorsatum
(Linnaeus)
Porcupine

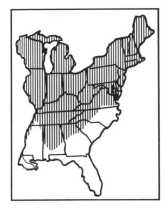

M. monax
(Linnaeus)
Woodchuck

G. sabrinus
(Shaw)
Northern Flying Squirrel

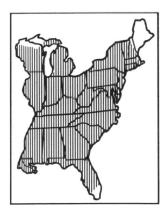

G. volans
(Linnaeus)
Southern Flying Squirrel

G. Four upper molariform teeth on each side of about equal size. Two dorsal light stripes. *Tamias striatus*

GG. Five upper molariform teeth, the first reduced in size. Four dorsal light stripes. . . . . *Eutamias minimus*

FF. Infraorbital opening a canal. Light stripes lacking or more than four

G. Zygoma not converging anteriorly, front portion not twisted toward the horizontal plane. Tail with great amount of hair and appears noticeably flattened

H. Anterior border of orbit ventrally opposite first large molariform teeth. Total length less than 400 mm . . . . . *Tamiasciurus hudsonicus*

HH. Anterior border of orbit ventrally opposite second large molariform teeth. Total length more than 400 mm . . . . *SCIURUS*

I. Four upper molariform teeth. Tail hairs yellow-tipped. . . . .*S. niger*

II. Five upper molariform teeth. Tail hairs white-tipped . *S. carolinensis*

GG. Zygoma converging anteriorly, front portion twisted toward the horizontal plane. Tail with less hair, and not flattened. . . . . . . *CITELLUS*

H. Animal with several light longitudinal stripes, and with light spots between the stripes . . . . *C. tridecemlineatus*

HH. No stripes. Animal brownish with obscure black flecks . . *C. franklini*

CC. Tail naked or nearly so. Skull without prominent postorbital processes

T. striatus
(Linnaeus)
Eastern Chipmunk

E. minimus
(Bachman)
Least Chipmunk

T. hudsonicus
(Erxleben)
Red Squirrel

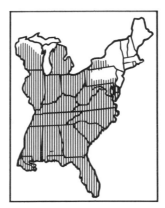

S. niger
Linnaeus
Eastern Fox Squirrel

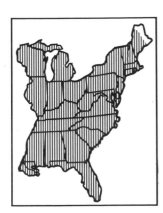

S. carolinensis
Gmelin
Eastern Gray Squirrel

C. tridecemlineatus
(Mitchill)
Thirteen-lined Ground
Squirrel

C. franklini
(Sabine)
Franklin's Ground
Squirrel

D. Skull very large. Incisors not grooved. Tail greatly flattened. No cheek pouches . . . . CASTORIDAE, *Castor canadensis*

DD. Skull smaller, with grooved incisors. Tail not greatly flattened. External fur-lined cheek pouches present . . . . GEOMYIDAE

E. Nasals not hour-glass shaped . . . . *Geomys bursarius*

EE. Nasals hour-glass shaped . . . .*G. pinetis* (includes *G. cumber-landius, G. colonus,* and *G. fontanelus* which are presently listed as species)

BB. Infraorbital opening larger, accomodating major muscles. Tail without a great deal of long fur and not flattened dorsoventrally. No external fur lined cheek pouches

C. Infraorbital opening oval (upper incisors with prominent groove). Tail very long, at least 1-1/3 times length of head and body . . . . ZAPODIDAE

D. Four upper molariform teeth, the first reduced in size. Tail normally without white tip . . . . *Zapus hudsonius*

DD. Three upper molariform teeth, essentially equal in size. Tail with white tip . . . .*Napaeozapus insignis*

CC. Infraorbital opening "V" shaped. Tail shorter

D. Molariform teeth with three longitudinal rows of cusps . . . . .MURIDAE

E. Total length less than 250 mm. Tail less than 110 mm. Skull less than 25 mm . . . . *Mus musculus*

EE. Total length more than 250. Tail more than 110. Skull more than 25 . . . . *RATTUS*

F. Tail longer than head and body. First upper molar with distinct outer notches on first row of cusps . . . .*R. rattus*

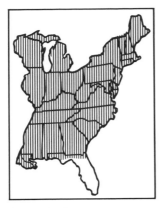

C. canadensis
Kuhl
Beaver

G. bursarius
(Shaw)
Prairie Pocket Gopher

G. pinetis
Rafinesque
Piney Woods Gopher

Z. hudsonius
(Zimmermann)
Meadow Jumping Mouse

N. insignis
(Miller)
Woodland Jumping Mouse

M. musculus
Linnaeus
House Mouse

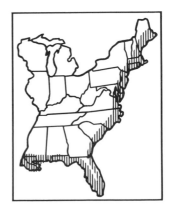

R. rattus
(Linnaeus)
Black Rat

FF. Tail not longer than head and body. First upper molars without such cusps . . . . *R. norvegicus*

DD. Molariform teeth with two rows of cusps, or without cusps . . . . CRICETIDAE

E. Cheek teeth with cusps (may be flattened into transverse lophs, or in the case of very old mice, may be worn flat then outlined with an even rim of enamel). Tail more than 1/3 of total length and well haired . . . . Subfamily CRICETINAE

F. Grooved upper incisors. Small species appearing much like a house mouse . . . . *REITHRODONTOMYS*

G. Tail usually more than 110% of body length. Last lower molar with dentine in the form of an "S" (only in Louisiana and SW Mississippi). . . . . *R. fulvescens*

GG. Tail usually less than 110% of body length; last lower molar with dentine in the form of a "C"

H. A distinct labial shelf or ridge, often with distinct cusplets on first and second lower molars . . . . *R. humulis*

HH. Without such a ridge (SW Wisc., and NW Ill.) . . . . *R. megalotis*

FF. Upper incisors not grooved. Not resembling a housemouse

G. Molars with prismatic pattern resembling rootless teeth. Rostrum very long and narrow. Animals large, at least 310 mm in length. White below and with large ears . . . . *Neotoma floridana*

GG. Molars with cusps or lophs. Rostrum not as long and narrow. Animals much smaller or else with grayish underparts.

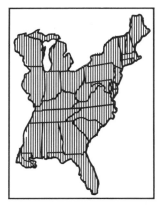

R. norvegicus
(Berkenout)
Norway Rat

R. fulvescens
Allen
Fulvous Harvest Mouse

R. humulis
(Audubon and Bachman)
Eastern Harvest Mouse

R. megalotis
(Baird)
Desert Harvest Mouse

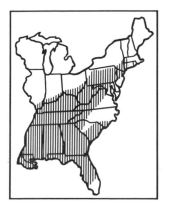

N. floridana
(Ord)
Florida Packrat

H. Zygoma deeply emarginate in front as seen from above because of shape and position of infraorbital foramen. Underparts usually grayish

  I. Molars with transverse lophs. Palate ends at end of molars. Fur very grizzled . . . . *Sigmodon hispidus*

  II. Molars with cusps. Fur much less grizzled, softer *Oryzomys palustris*

HH. Zygoma not deeply emarginate. Molars with cusps. Underparts and feet white

  I. Plantar pads five . . . . *Peromyscus floridanus*

  II. Plantar pads six

    J. Ears bright ochraceous, same color as body. Posterior palatine foramina nearer to interpterygoid fossa than to anterior palatine foramina . . . . *Ochrotomys nuttalli*

    JJ. Ears dusky, contrasting slightly with body color. Posterior palatine foramina midway between openings mentioned above . . . . *PEROMYSCUS*

      K. Very light colored, and small in size (tail usually less than 55 mm, and hind foot usually less than 18 mm) . . . . . *P. polionotus*

      KK. Darker in color, usually larger

        L. Grayish in color dorsally, with many black hairs. Dorsal stripe indistinct or absent . . . . *P. maniculatus*

        LL. Brownish color, with broad dorsal stripe generally fairly well defined in adults, fur softer, with less black hairs interspersed

S. hispidus
Say and Ord
Common Cotton Rat

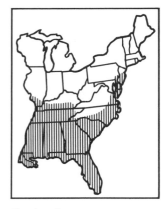

O. palustris
(Harlan)
Eastern Rice Rat

P. floridanus
(Chapman)
Gopher Mouse

O. nuttalli
(Harlan)
Golden Mouse

P. polionotus
(Wagner)
Beach Mouse

P. maniculatus
(Wagner)
Deer Mouse

M. Hind foot usually more than 22 mm, southern species usually inhabiting lowlands . . . . *P. gossypinus*

MM. Hind foot usually less than 22 mm . . . . *P. leucopus*

EE. Cheek teeth with loops and triangles; rootless. Tail short, less than 1/3 of total length, or if long, then animal large, at least 285 mm, and with tail only scantily haired . . . . Subfamily MICROTINAE

F. Animal large, at least 285 mm in total length. Tail scantily haired, basal length of skull at least 40 mm. Postorbital processes nearly right-angled and projecting into orbit

G. Tail round. Basal length of skull less than 50 mm . . . . . *Neofiber alleni*

GG. Tail laterally flattened. Basal length of skull more than 50 mm. . . . . *Ondatra zibethicus*

FF. Animal less than 285 mm. Tail well haired. Basal length of skull less than 40 mm

G. Palate ending in straight shelf; color of animal generally reddish (some exceptions). . . . . *Clethrionomys gapperi*

GG. Palate not ending in straight shelf. Color not generally reddish (except sometimes in *Microtus pinetorum*, then the fur is very soft and has a tendency to lie either backwards or forwards)

H. Tail about equal to length of hind foot. Fur coarse. Incisors with longitudinal groove . . . . SYNAPTOMYS

P. gossypinus
(Le Conte)
Cotton Mouse

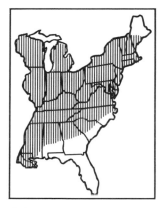

P. leucopus
(Rafinesque)
Wood Mouse

N. alleni
True
Florida Water Rat

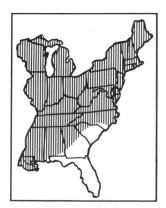

O. zibethicus
(Linnaeus)
Muskrat

C. gapperi
(Vigors)
Red-backed Vole

I. None of the hairs at base of ears appreciably brighter than remainder of pelage. Lower molars with triangles on outer sides; palate with broad, blunt median projection. . . . . . *Synaptomys cooperi*

II. A few hairs at base of ears distinctly rust colored. Lower molars without triangles on outer side; palate with sharp-pointed, median projection . . . . *S. borealis*

HH. Tail longer than hind foot, or else fur very fine. No grooves in incisors . . . . *MICROTUS*

I. Tail about equal to hind foot, fur very fine, tending to lie either way (skull very similar to that of *M. ochrogaster*). . . . . *M. pinetorum*

II. Tail longer than hind foot. Fur coarse

J. Third upper molar with 5 closed triangles. Yellowish colored nose (species of Allegheny Mountain Region) . . . . . *M. chrotorrhinus*

JJ. Third upper molar with 2 or 3 closed triangles. Nose not yellowish

K. Third upper molar with 3 closed triangles. Ventral fur silvery. Tail usually much more than twice the length of the hind foot . . . . *M. pennsylvanicus*

KK. Third upper molar with 2 closed triangles. Ventral fur buff colored. Tail usually about twice length of hind foot . . . . *M. ochrogaster*

S. cooperi
Baird
Southern Bog Lemming

S. borealis
(Richardson)
Northern Bog Lemming

M. pinetorum
(Le Conte)
Pine Vole

M. chrotorrhinus
(Miller)
Yellow-cheeked Vole

M. pennsylvanicus
(Ord)
Meadow Vole

M. ochrogaster
(Wagner)
Prairie Vole

BIBLIOGRAPHY

Listed here are a group of papers and books which might prove of interest to individuals using these keys.

FISHES.

Abbott, C. C. 1870. Notes on fresh-water fishes of New Jersey. Am. Nat. 4:99-117.

_____. 1874. Notes on the Cyprinoids of central New Jersey. Am. Nat. 8:326-338.

Adams, C., and T. L. Hankinson. 1928. The ecology and economics of Oneida Lake fish. Bull. N. Y. St. Coll. Forestry, Syracuse Univ. 1:239-548.

Agassiz, L. 1854. Notice on a collection of fishes from the southern bend of the Tennessee River, Alabama. Am. Jour. Sci. and Arts 17:297-308, 353-365.

American Fisheries Society. 1960. A list of common and scientific names of fishes from the United States and Canada. Am. Fish. Soc. Spec. Publ. 2. 102 p.

Bailey, J. R. and D. G. Frey. 1951. Darters of the genus *Hololepis* from some natural lakes of North Carolina. J. Elisha Mitchell Sci. Soc. 67:191-204.

Bailey, J. R., and J. A. Oliver. 1939. The fishes of the Connecticut watershed. New Hamp. Fish and Game Dept. Surv. Rept. 4:150-189.

Bailey, R. M. 1938. The fishes of the Merrimack watershed. Biol. Surv. Merrimack Watershed. New Hamp. Fish and Game Dept., Surv. Rep. 3:149-185.

_____. 1959. *Etheostoma acuticeps*, a new darter from the Tennessee River system, with remarks on the subgenus *Nothonotus*. Occ. Pap. Mus. Zool. Univ. Mich. 603:1-10.

_____. 1959. A new catostomid fish, *Moxostoma (Thoburnia) atripinne*, from the Green River drainage, Kentucky and Tennessee. Occ. Pap. Mus. Zool. Univ. Michigan. 599. 19 p.

Bailey, R. M. and F. B. Cross. 1954. River sturgeons of the American genus *Scaphirhynchus*: Characters, distribution, and synonymy. Pap. Mich. Acad. Sci. Arts and Letters, 39(1953):169-208.

Bailey, R. M. and W. A. Gosline. 1955. Variation and systematic significance of vertebral counts in the American fishes of the family Percidae. Misc. Publ. Mus. Zool. Univ. Michigan 93:5-44.

Bailey, R. M. and C. L. Hubbs. 1949. The Black Basses *(Micropterus)* of Florida, with description of a new species. Occas. Pap. Mus. Zool. Univ. Mich., 516. 40 p.

Bailey, R. M., and W. J. Richards. 1963. Status of *Poecilichthys hopkinsi* Fowler and *Etheostoma trisella*, new species, Percid fishes from Alabama, Georgia, and South Carolina. Occ. Pap. Mus. of Zool. Univ. Mich. 630:1-21.

Bailey, R. M., H. E. Winn and C. L. Smith. 1954. Fishes from the Escambia River, Alabama and Florida, with ecologic and taxonomic notes. Proc. Acad. Nat. Sci. Phil. 106:109-164.

Baker, C. L. Key to Reelfoot Lake fishes. J. Tenn. Acad. Sci. 14:41-45.

_____. 1939. Additional fishes of Reelfoot Lake. J. Tenn. Acad. Sci. 14:6-40.

Bean, B. A. 1903. Notice on a collection of fishes made by H. H. Brimley in Cane River and Bollings Creek, North Carolina, with a description of new species of *Notropis (N. brimleyi)*. Proc. U. S. Nat. Mus. 26(1339):913-914.

Bean, H. 1903. Catalogue of the fishes of New York. N. Y. St. Mus. Bull. 60. 784 p.

Bean, T. H. 1892. The fishes of Pennsylvania, with descriptions of the species and notes of their common names, distribution, habits, reproduction, rate of growth and mode of capture. Rept. (Pa.) State Comm. Fish. (1889-91): 1-149.

Behnke, R. J. and R. M. Wetzel. 1960. A preliminary list of the fishes found in the fresh waters of Connecticut. Copeia 1960: 141-143.

228

Berg, L. S. 1947. Classification of fishes both recent and fossil. J. W. Edwards, Ann Arbor. 517 p.

Blatchley, W. S. 1938. The fishes of Indiana. The Nature Publ. Co., Indianapolis. 121 p.

Briggs, J. C. 1958. A list of Florida fishes and their distribution. Bull. Florida State Mus. 2:223-318.

Cahn, Alvin R. 1927. An ecological study of southern Wisconsin Fishes. The brook Silverside *(Labidesthes sicculus)* and the Cisco *(Leucichthys artedi)* in their relations to the region. Ill. Biol. Monogr., 11:1-151.

Call, R. E. 1896. Fishes and shells of the falls of the Ohio. Mem. Hist. Louisville 1:9-20.

Carpenter, R. G. and H. R. Siegler. 1947. Fishes of New Hampshire. New Hamp. Fish and Game Comm. 87 p.

Carr, A. 1937. A Key to the fresh water fishes of Florida. Proc. Fla. Acad. Sci. 1:72-86.

Carr, A. F., Jr., and C. J. Goin. 1955. Guide to the reptiles, amphibians, and fresh-water fishes of Florida. Univ. Fla. Press, Gainesville. 341 p.

Coker, E. 1930. Studies of common fishes of the Mississippi River at Keokuk. Bull. U. S. Bur. Fish. 45:141-225.

Cole, C. F. 1965. Additional evidence for separation of *Etheostoma olmstedi* Storer from *Etheostoma nigrum* Rafinesque. Copeia 1965:8-13.

Collette, B. B. 1962. The swamp darters of the subgenus *Hololepis* (Pisces, Percidae). Tul. Stud. Zool. 9:115-211.

_____. 1963. The subfamilies, tribes, and genera of the Percidae (Teleostei). Copeia 1963:615-623.

_____. 1965. Systematic significance of breeding tubercles in fishes of the family Percidae. Proc. U. S. Nat. Mus. 117:567-614.

Collette, B. B., and R. W. Yerger. 1962. The American Percid fishes of the subgenus Villora. Tul. Stud. Zool. 9:213-230.

Cook, F. A. 1959. Freshwater fishes in Mississippi. Miss. Game and Fish. Comm. 239 p.

Cooper, G. P. 1939. A biological survey of the waters of York County and the southern part of Cumberland County, Maine. Maine Dept. Inland Fish. and Game, Fish Surv. Rept. 1:1-58.

_____. 1939. A biological survey of thirty-one lakes and ponds of the Upper Saco River and Sebago Lake Drainage Systems in Maine. Maine Dept. Inland Fish. and Game, Fish Surv. Rept. No. 2. 147 p.

_____. 1940. A biological survey of the Rangeley Lakes, with special reference to trout and salmon. Maine Dept. Inland Fish. and Game, Fish Surv. Rept. No. 3. 182 p.

_____. 1941. A biological survey of lakes and ponds of the Androscoggin and Kennebec River drainage systems in Maine. Maine Dept. Inland Fish. and Game, Fish Surv. Rept. No. 4. 238 p.

_____. 1942. A biological survey of lakes and ponds of the central coastal area of Maine. Maine Dept. of Inland Fish. and Game, Fish Surv. Rept. No. 5. 184 p.

Cooper, G. P., and J. L. Fuller. 1945. A biological survey of Moosehead Lake and Haymock Lake, Maine. Maine Dept. Inland Fish. and Game, Fish Surv. Rept. No. 6. 160 p.

Cope, E. D. 1864. Partial catalogue of the cold-blooded vertebrata of Michigan, Pt. I. Proc. Acad. Nat. Sci. Phil. 16:276-285.

_____. 1869. On the distribution of fresh-water fish in the Allegheny region of southwestern Virginia. Jour. Acad. Nat. Sci. Phil., Ser. 2, 6:207-247.

_____. 1870. On some etheostomine perch from Tennessee and North Carolina. Proc. Amer. Philos. Soc. 11:261-270.

_____. 1870. A partial synopsis of the fishes of the fresh waters of North Carolina. Trans-Amer. Phil. Soc. 11:448-495.

_____. 1881. The fishes of Pennsylvania. Rept. (Pa.) State Comm. Fish. (1879-80): 59-145.

Creaser, W. 1926. The structure and growth of the scales of fishes in relation to the interpretation of their life-history, with special reference to the sunfish, *Eupomotis gibbosus*. Misc. Publ. Mus. Zool. Univ. Mich., No. 17, 82 p.

Crossman, E. J. 1962. The redfin pickerel, *Esox a. americanus* in North Carolina. Copeia 1962:114-123.

_____. 1966. A taxonomic study of *Esox americanus* and its subspecies in Eastern North America. Copeia 1966:1-20.

Curtis, B. 1938. The life story of the fish. Appleton-Century. N. Y. 260 p.

Dymond, R. 1922. A provisional list of the fishes of Lake Erie. Publ. Ont. Fish. Res. Lab. 4:55-73.

Eddy, S. 1957. How to know the freshwater fishes. Wm. C. Brown Co., Dubuque. 253 p.

Eddy, S. and A. C. Hodson. 1961. Taxonomic Keys to the common animals of the north central states. Burgess Publishing Co. 166 p.

Eddy, S. and T. Surber. 1947. Northern fishes with special reference to the upper Mississippi Valley. 2nd Ed. Univ. Minn. Press, Minneapolis. 276 p.

Eigenmann, C. H. and C. H. Beeson. 1894. Fishes of Indiana. Proc. Inc. Acad. Sci. 3:76-108.

Elser, H. J. 1950. The common fishes of Maryland: How to tell them apart. Md. Bd. Nat. Res. Dept. Res. and Educ. 88:1-45.

Everhart, W. H. 1950. Fishes of Maine. Maine Dept. Inland Fish and Game. 1950: 1-53.

Evermann, B. W. 1902. List of species of fishes known to occur in the Great Lakes or their connecting waters. Bull. of U. S. Fish Comm. 21:95-96.

_____. 1918. The fishes of Kentucky and Tennessee: A distributional catalogue of the known species. Bull. U. S. Bur. Fish. 35:295-368.

Evermann, B. W. and H. W. Clark. 1920. Lake Maxinkuckee. A physical and biological survey. Ind. Dept. Cons. Vol. 1, 660 p.

_____. 1930. Check list of the fishes and fishlike vertebrates of North and Middle America north of the northern boundary of Venezuela and Colombia. Rept. U. S. Comm. Fish. 1928(2):1-670.

Evermann, B. W., and U. O. Cox. 1896. The fishes of the Neuse River basin. Bull. U. S. Fish Comm. 15:303-310

Evermann, B. W. and S. F. Hildebrand. 1916. Notes on the fishes of east Tennessee. Bull. U. S. Bur. Fish. 34:431-451.

Evermann, B. W. and W. C. Kendall. 1900. Check list of the fishes of Florida. Rept. U. S. Fish Comm. 1899:37-103.

_____. 1902. Notes on the fishes of Lake Ontario. Rept. U. S. Fish Comm. 21:209-216.

_____. 1902. An annotated list of the fishes known to occur in Lake Champlain and its tributary waters. Rept. U. S. Fish Comm. 1901:217-225.

_____. 1902. An annotated list of the fishes known to occur in the St. Lawrence River. Rept. U. S. Fish Comm. 1901:227-240.

Evermann, B. W., and H. B. Latimer. 1910. The fishes of the Lake of the Woods and Connection waters. Proc. U. S. Nat. Mus. 39:121-136.

Fish, M. P. 1932. Contributions to the early life histories of sixty-two species of fishes from Lake Erie and its tributary waters. Bull. U. S. Bur. Fish. 47:293-398.

Forbes, S. A. 1907. On the local distribution of certain Illinois fishes: an essay in statistical ecology. Bull. Ill. State Lab. Nat. Hist. 7:273-303.

_____. 1909. On the general and interior distribution of Illinois fishes. Bull. Ill. State Lab. Nat. Hist. 8:381-437.

_____. 1914. Fresh water fishes and their ecology. Ill. State Lab. Nat. Hist. Urbana. 19 p.

230

Forbes, S. A., and R. E. Richardson. 1920. The fishes of Illinois. 2nd Ed. Ill. Nat. Hist. Surv., Springfield. 358 p.

Fowler, H. W. 1906. The fishes of New Jersey. Ann. Rept. N. J. State Mus. 1905(2): 35-477.

_____. 1907. A supplementary account of the fishes of New Jersey. Ann. Rept. N. J. State Mus. 1906(3):251-384.

_____. 1909. A synopsis of the Cyprinidae of Pennsylvania. Proc. Acad. Nat. Sci. Phil. 60:517-553.

_____. 1911. The fishes of Delaware. Proc. Acad. Nat. Sci. Phil. 63:3-16.

_____. 1918. A review of the fishes described in Cope's partial catalogue of the cold-blooded vertebrata of Michigan. Occ. Papers Mus. Zool. Univ. Mich. 60:1-51.

_____. 1923. Records of fishes for the eastern and southern United States. Proc. Acad. Natur. Sci. Phil. 74:1-27.

_____. 1923. Records of fishes for the southern states. Proc. Biol. Soc. Wash. 36: 7-34.

_____. 1935. Notes on South Carolina fresh-water fishes. Contr. Charleston Mus. 7: 1-28.

_____. 1936. Freshwater fishes obtained in North Carolina in 1930 and 1934. Fish Culturist 15:192-194.

_____. 1940. A list of the fishes recorded from Pennsylvania. Bull. Commonwealth of Pennsylvania Bd. Fish. 7:3-25.

_____. 1941. A collection of fresh-water fishes obtained in Florida, 1939-40, by Francis Harper. Proc. Acad. Nat. Sci. Phil. 92:227-244.

_____. 1945. A study of the fishes of the southern Piedmont and coastal plain. Monogr. Acad. Nat. Sci. Phil. 7:1-408.

_____. 1948. A list of the fishes recorded from Pennsylvania. Bull. Board of (Pa.) Fish Comm. 7:3-26.

_____. 1952. A list of the fishes of New Jersey, with off-shore species. Proc. Acad. Nat. Sci. 104:89-151.

Freeman, H. W. 1952. Fishes of Richland County, South Carolina. Univ. South Carolina Publ. Biol., Ser. III, 1:28-41.

_____. 1952. New distribution records for fishes of the Savannah River basin, South Carolina. Copeia 1952:269.

_____. 1954. Fishes of the Savannah River operations area. Univ. South Carolina Publ. Biol., Ser. III, 1:117-156.

Frey, D. G. 1951. The fishes of North Carolina's Bay Lakes and their intraspecific variation. J. Elisha Mitchell Sci. Soc. 67:1-44.

Gage, S. H. 1928. The lampreys of New York State - life history and economics. Biol. Surv. of the Oswego River System. Suppl. 17th Ann. Rept. N. Y. St. Cons. Dept. 158-191.

_____. 1929. Lampreys and their ways. Sci. Monthly, 28:401-416.

Gerking, S. D. 1945. The distribution of the fishes of Indiana. Invest. Ind. Lakes and Streams 3:1-137.

_____. 1955. Key to the fishes of Indiana. Invest. Ind. Lakes and Streams 4:49-86.

Gibbs, R. H., Jr. 1957. Cyprinid fishes of the subgenus Cyprinella of Notropis. I. Systematic status of the subgenus Cyprinella, with a key to the species exclusive of the lutrensis-ornatus complex. Copeia 1957:185-195.

Gilbert, C. H. 1844. A list of fishes collected in the East Fork of White River, Indiana, with descriptions of two new species. Proc. U. S. Nat. Mus. 7(423):199-205.

_____. 1884. Notes on the fishes of the Switz City Swamp, Greene County, Indiana. Proc. U. S. Mus. 7:206-210.

_____. 1891. Report of the explorations made in Alabama during 1889, with notes on the fishes of the Tennessee, Alabama, and Escambia rivers. Bull. U. S. Fish Comm. 9:143-159.

Gilbert, C. R. 1964. The American cyprinid fishes of the subgenus Luxilus (Genus Notropis). Bull. Fla. State Mus. 8:95-194.

Gill, T. N. 1904. A remarkable genus of fishes - the Umbras. Smiths. Misc. Coll. 45:295-305.

Goldsborough, E. L. and H. W. Clark. 1908. Fishes of West Virginia. Bull U. S. Bur. Fisheries for 1907, 27:29-39.

Gordon, M. 1937. The fishes of eastern New Hampshire. Biol. Surv. Androscoggin, Saco and Coastal Watersheds. New Hamp. Fish and Game Dept., Surv. Rep. 2: 101-118.

Gosline, W. A. 1948. Speciation in the fishes of the genus *Menidia*. Evolution 2:306-313.

Greeley, J. R. 1927. Fishes of the Genesee region with annotated list. In: A biological survey of the Oswego River system. Suppl. 16th Ann. Rept. N. Y. State. Cons. Dept. 1926:47-66.

_____. 1928. Fishes of the Oswego watershed with annotated list. In: A biological survey of the Oswego River system. Suppl. 17th Ann. Rept. N. Y. State Cons. Dept. 1927:84-107.

_____. 1929. Fishes of the Erie-Niagara watershed with annotated list, In: A biological survey of the Erie-Niagara system. Suppl. 18th Ann. Rept. N. Y. State Cons. Dept. 1929:150-179.

_____. 1930. Fishes of the Lake Champlain watershed with annotated list, In: A biological survey of the Champlain watershed. Suppl. 19th Ann. Rept. N. Y. State Cons. Dept. 1929:44-87.

_____. 1934. Annotated list of the fishes occurring in the watershed, In: A biological survey of the Raquette watershed. Suppl. 23rd Ann. Rept. N. Y. State Cons. Dept. 1933:53-108.

_____. 1935. Fishes of the watershed with annotated list, In: A biological survey of the Mohawk-Hudson watershed. Suppl. 24th Ann. Rept. N. Y. State Cons. Dept. 1934: 63-101.

_____. 1936. Fishes of the area with annotated list, In: A biological survey of the Delaware and Susquehanna watershed. Suppl. 25th Ann. Rept. N. Y. State Cons. Dept., 1935, p. 45-88.

_____. 1937. Fishes of the area with annotated list, In: A biological survey of the lower Hudson watershed. Suppl. 26th Ann. Rept. N. Y. State Cons. Dept. 1936: 45-103.

_____. 1938. Fishes of the area with annotated list, In: A biological survey of the Allegheny and Chemung watersheds. Suppl. 27th Ann Rept. N. Y. State Cons. Dept. 1937:48-73.

_____. 1939. The freshwater fishes of Long Island and Staten Island with annotated list, In: A biological survey of the fresh waters of Long Island. Suppl. 28th Ann. Rept. N. Y. State Cons. Dept. 1938:29-44.

_____. 1940. Fishes of the watershed with annotated list, In: A biological survey of the Lake Ontario watershed. Suppl. 29th Ann. Rept. N. Y. State Cons. Dept. 1939: 42-81.

Greeley, J. R. and S. C. Bishop. 1932. Fishes of the area with annotated list, In: A biological survey of the Oswegatchie and Black River systems. Suppl. 21st Ann. Rept. N. Y. State Cons. Dept. 1931:54-92.

_____. 1933. Fishes of the upper Hudson watershed with annotated list, In: A biological survey of the upper Hudson watershed. Suppl. 22nd Ann. Rept. N. Y. State Cons. Dept. 1932:64-101.

Greeley, J. R. and C. W. Greene. 1931. Fishes of the area with annotated list, In: A biological survey of the St. Lawrence watershed. Suppl. 20th Ann. Rept. N. Y. State Cons. Dept. 1930:44-94.

Greene, C. W. 1935. The distribution of Wisconsin fishes. Wisc. Cons. Comm. Madison. 235 p.

Gunning, G. E. and W. M. Lewis. 1955. The fish population of a spring-fed swamp in the Mississippi bottoms of southern Illinois. Ecology 36:552-558.

Hankinson, T. L. 1932. Observations on the breeding behavior and habits of fishes in southern Michigan. Papers Mich. Acad. Sci., Arts and Letters 15:411-425.

Hay, O. P. 1894. The lampreys and fishes of Indiana. 19th Ann. Rept. Ind. Dept. Geol. and Nat. Res:147-296.

Henshall, J. A. 1919. Bass, pike and perch and other game fishes of America. Stewart and Kidd, Cincinnati. 410 p.

Herald, E. S. 1961. Living fishes of the world. Doubleday. N.Y. 304 p.

Hildebrand, S. F. 1932. On a collection of fishes from the Tuckaseegee and upper Catawba river basins, N. C., with a description of a new darter. Jour. Elisha Mitchell Sci. Soc. 48:50-82.

Hubbs, C. L. 1925. The life-cycle and growth of lampreys. Papers Mich. Acad. Sci., Arts and Letters 4:587-603.

_____. 1930. Materials for a revision of the Catostomid fishes of eastern North America. Misc. Publ. Mus. Zool. Univ. Mich. 20:1-47.

_____. 1945. Corrected distributional records for Minnesota fishes. Copeia 1945:13-22.

Hubbs, C. L., and E. R. Allen. 1944. Fishes of Silver Springs, Florida. Proc. Fla. Acad. Sci. 6:110-130.

Hubbs, C. L. and R. M. Bailey. 1940 A revision of the Black Basses (*Micropterus* and *Huro*) with descriptions of four new forms. Misc. Publ. Mus. Zool. Univ. Mich. 48:1-51.

Hubbs, C. L. and J. D. Black. 1940. Percid fishes related to *Poecilichthys variatus*, with descriptions of three new forms. Occas. Pap. Mus. Zool. Univ. Mich. 416:1-30.

Hubbs, C. L., and D. E. S. Brown. 1929. Materials for a distributional study of Ontario fishes. Trans. Roy. Can. Inst. Part I. 18:1-56.

Hubbs, C. L., and M. D. Cannon. 1935. The darters of the genera *Hololepis* and *Villora*. Misc. Publ. Mus. Zool. Univ. Mich. 30:1-93.

Hubbs, C. L., and G. P. Cooper. 1936. The minnows of Michigan, with special reference to ten common species. Bull. Cranbrook Inst. Sci. 8:95 p.

Hubbs, C. L., and W. R. Crowe. 1956. Preliminary analysis of the American Cyprinid fishes, seven new, referred to the genus *Hybopsis*, Subgenus *Erimystax*. Occas. Pap. Mus. Zool. Univ. Mich. 578:1-8.

Hubbs, C. L., and C. W. Greene. 1928. Further notes on the fishes of the Great Lakes and tributary waters. Pap. Mich. Acad. Sci., Arts and Letters 8:371-392.

Hubbs, C. L., L. C. Hubbs, and R. E. Johnson. 1943. Hybridization in nature between species of Catostomid fishes. Contr. Lab. Vert. Biol. Univ. Mich. 22:1-76.

Hubbs, C. L., and K. F. Lagler. 1943. Annotated list of the fishes of Foots Pond, Gibson County, Indiana. Invest. Ind. Lakes and Streams 2:73-83.

_____. 1958. Fishes of the Great Lakes Region. Bull. Cranbrook Inst. Sci. 26. 213 p.

Hubbs, C. L., and E. C. Raney. 1944. Systematic notes on North American Siluroid fishes of the Genus *Schilbeodes*. Occas. Pap. Mus. Zool. Univ. Mich. 487:1-37.

_____. 1946. Endemic fish fauna of Lake Waccamaw, North Carolina. Misc. Publ. Univ. Mich. Mus. Zool. 65:1-30.

Jordan, D. S. 1877. Partial synopsis of the fishes of upper Georgia; with supplementary papers on fishes of Tennessee, Kentucky, and Indiana. Ann. N. Y. Lyceum Nat. Hist. 11:307-377.

_____. 1877. On the fishes of northern Indiana. Proc. Acad. Natur. Sci. Phil. 29:42-82.

_____. 1890. Report of explorations made during 1888 in the Allegheny region of Virginia, North Carolina, and Tennessee, and in western Indiana with an account of the fishes found in each of the river basins of these regions. Bull. U. S. Fish. Comm. 8:97-173.

_____. 1929. Manual of the vertebrate animals of the northeastern United States. 13th ed. World Book Company, Yonkers, N. Y. 446 p.

Jordan, D. S., and H. E. Copeland. 1877. Check list of the fishes of fresh waters of North America. Bull. Buffalo Soc. Natur. Sci. 3:133-164.

Jordan, D. S., and B. W. Evermann. 1896. The fishes of north and middle America. Bull. U. S. Nat. Mus. 47:1-1240.

_____. 1902. American food and game fishes. A popular account of all the species found in America north of the equator, with keys for ready identification, life histories and methods of capture. Doubleday, Page and Co., N. Y. 573 p.

Jordan, D. S., B. W. Evermann, and H. W. Clark. 1930. Checklist of the fishes and fishlike vertebrates of North and Middle America, north of the northern boundary of Venezuela and Colombia. Rept. U. S. Comm. Fish. 1928. 670 p.

Jordan, D. S., and C. H. Gilbert. 1883. A synopsis of the fishes of North America. Bull. U. S. Natl. Mus. 16:1-1018 p.

Jordan, D. S., and J. Swain. 1883. List of fishes collected in the Clear Fork of the Cumberland, Whitley County, Kentucky, with descriptions of three new species. Proc. U. S. Nat. Mus. 6(378):248-251.

Kendall, W. C. 1908. Fauna of New England. 8. List of the Pisces. Occas. Pap. Boston Soc. Nat. Hist. 7:1-152.

_____. 1910. American catfishes: Habits, culture and commercial importance. Rept. U. S. Comm. Fish for 1908. 39 p.

_____. 1914. The fishes of New England: The salmon family. 1. The trout or charrs. Mem. Boston Soc. Nat. Hist. 8:1-103.

_____. 1935. The fishes of New England: The Salmon Family. 2. The salmons. Mem. Boston Soc. Nat. Hist. 9:1-166.

Kendall, W. C., and W. A. Deuce. 1929. The fishes of the Cranberry Lake Region. Roosevelt Wild Life Bull. 5:219-309.

King, W. 1947. Important food and game fishes of North Carolina. Dept. Cons. and Devel. Div. Game and Inland Fish. 54 p.

Knapp, L. W., W. J. Richards, R. V. Miller, and N. R. Foster. 1963. Rediscovery of the Percid fish *Etheostoma sellare* (Radcliffe and Welch). Copeia 1963:455.

Koelz, W. 1929. Coregonid fishes of the Great Lakes. Bull. U. S. Bur. Fish. 43(2):297-643.

_____. 1931. The Coregonid fishes of northeastern North America. Pap. Mich. Acad. Sci., Arts and Letters 13(1930):303-432.

Kuhne, E. R. 1939. A guide to the fishes of Tennessee and the mid-South. Div. Fish and Game, Tenn. Dept. Cons. Nashville. 124 p.

Kyle, H. M. 1926. The biology of fishes. Macmillan. N. Y. 396 p.

Lachner, E. A., E. F. Westlake and P. S. Handwerk. 1950. Studies of the biology of some percid fishes from western Pennsylvania. Amer. Midland Nat. 43:92-111.

Lachner, E. A. 1952. Studies of the biology of the Cyprinid fishes of the chub genus *Nocomis* of northeastern United States. Amer. Midland Nat. 48:433-466.

LaMonte, F. 1945. North American game fishes. Doubleday, Garden City. 202 p.

Larimore, R. W. and P. W. Smith. 1963. The fishes of Champaign County, Illinois, as affected by 60 years of stream changes. Ill. Nat. Hist. Surv. Bull. 28:299-382.

Lennon, R. E. 1962. An annotated list of the fishes of the Great Smoky Mountains National Park. J. Tenn. Acad. Sci. 37:5-7.

Luce, W. M. 1933. A survey of the fishery of the Kaskaskia River. Bull. Ill. Nat. Hist. Surv. 20:75-123.

McConnell, W. R. 1906. Preliminary report of the investigation of certain waters of Pennsylvania. Rept. Dept. of Fish. Pa. 1905:172-179.

Meehean, O. L. 1942. Fish populations of five Florida lakes. Trans. Am. Fish. Soc. 71:184-194.

Meek, S. E. 1908. List of fishes known to occur in the waters of Indiana. Bien. Rept. Comm. Fish. and Game for Indiana:134-171.

234

Michael, E. L. 1906. A catalogue of Michigan fish. Bull. Mich. Fish. Comm. 8:1-45.

Nikolsky, G. V. 1963. The ecology of fishes. Acad. Press. N. Y. 352 p.

Miller, R. J. 1964. Behavior and ecology of some North American Cyprinid fishes. Amer. Midl. Nat. 72:313-357.

Miller, R. R. 1960. Systematics and biology of the gizzard shad *(Dorosoma cepedianum)* and related fishes. U. S. Fish and Wildlife Service Fishery Bull. 60:371-392.

Moore, G. A. 1957. Part 2. Fishes. pg. 31-210. In: Blair, *et al.*, Vertebrates of the United States. McGraw-Hill. N. Y. 819 p.

Nelson, E. W. 1876. A partial catalogue of the fishes of Illinois. Bull. Ill. Mus. Natur. His. 1:33-52.

Norman, J. R. 1936. A history of fishes. Ernest Benn Limited, London. 463 p.

O'Donnell, J. D. 1935. Annotated list of the fishes of Illinois. Ill. Nat. Hist. Surv. Bull. 20:473-500.

Osburn, R. C. 1901. The fishes of Ohio. Ohio State Acad. Sci. Spec. Pap. 4:5-105.

Osburn, R. C., E. L. Wickliff, and M. B. Trautman. 1930. A revised list of the fishes of Ohio. Ohio J. Sci. 30:169-176.

Pickens, A. L. 1928. Fishes of upper South Carolina. Copeia 167:29-32.

Raney, E. C. 1950. Freshwater fishes. In: The James River Basin past, present and future. Virginia Acad. of Sci., Richmond:151-194.

Raney, E. C. and R. D. Suttkus. 1964. *Etheostoma moorei,* a new darter of the subgenus *Nothonotus* from the White River system, Arkansas. Copeia 1964:131-139.

Raney, E. C. and T. Zorach. 1967. *Etheostoma microlepidum,* a new percid fish of the subgenus *Nothonotus* from the Cumberland and Tennessee River systems. Amer. Midl. Nat. 77:93-104.

Reid, G. K., Jr. 1950. The fishes of Orange Lake, Florida. Quart. J. Fla. Acad. Sci. 12:173-183.

Reighard, Jacob. 1915. An ecological reconnaissance of the fishes of Douglas Lake, Cheboygan County, Michigan, in midsummer. Bull. U. S. Bur. Fish. 33:215-249.

Richards, W. J., and L. W. Knapp. 1964. *Percina lenticula,* a new percid fish, with a redescription of the subgenus *Hadropterus.* Copeia 1964:690-701.

Robins, C. R. 1961. Two new cottid fishes from the fresh waters of eastern United States. Copeia 1961:305-315.

Robins, C. R., and E. C. Raney. 1956. Studies of the Catostomid fishes of the genus *Moxostoma,* with descriptions of two new species. Cornell Univ. Mem. 343. 56 p.

Ross, R. D. 1959. Drainage evolution and distribution problems of the fishes of the New (Upper Kanawha) River system in Virginia. IV. Key to the identification of fishes. Virginia Agr. Exp. Sta. Tech. Bull. 146:3-27.

Schrenkeisen, R. 1938. Field book of freshwater fishes of North America north of Mexico. G. P. Putnam's Sons, N. Y. 312 p.

Schwartz, F. J. and J. Norvell. 1958. Food, growth and sexual dimorphism of the redside dace *Clinostomus elongatus* (Kirtland) in Linesville Creek, Crawford County, Pennsylvania. Ohio Jour. Sci. 58:311-316.

Shoemaker, H. H. 1942. The fishes of Wayne County, Indiana. Invest. Ind. Lakes and Streams. 2:267-296.

Shoup, C. S., J. H. Peyton and G. Gentry. 1941. A limited biological survey of the Obey River and adjacent streams in Tennessee. Rep. Reelfoot Lake Biol. Sta. 5: 48-76.

Smith, H. M. The fishes of North Carolina. N. C. Geol. and Econ. Surv. 2:1-449.

Smith, R. F. 1953. Some observations on the distribution of fishes in New Jersey. N. J. Fish. Surv., Div. Fish and Game Rep. 2:165-174.

Taylor, W. R. 1954. Records of fishes in the John N. Lowe collection from the Upper Peninsula of Michigan. Misc. Publ. Mus. Zool., Univ. Mich. 87:1-50.

Thompson, D. H., and F. D. Hunt. 1930. The fishes of Champaign County: A study of the distribution and abundance of fishes in small streams. Bull. Ill. State Nat. Hist. Surv. 19:1-101.

Trautman, M. B. 1957. The fishes of Ohio. Ohio State Univ. Press. Columbus. 683 p.

Truitt, R. V., B. A. Bean, and H. W. Fowler. 1929. The fishes of Maryland. Md. Cons. Dept. Bull. 3:1-120.

Webster, D. A. 1942. The life histories of some Connecticut fishes. In: A fishery survey of important Connecticut lakes. Conn. St. Geol. and Nat. Hist. Surv. Bull. 63.

Weed, A. C. 1925. A review of the fishes of the genus *Signalosa*. Field Mus. Nat. Hist. Publ. 233, 12(11): 137-146.

Welter, W. A. 1938. A list of the fishes of the Licking River drainage in eastern Kentucky. Copeia 1938:64-68.

Wickliff, E. L., and M. B. Trautman. 1932. Some food and game fishes of Ohio. Bull. Ohio Dept. Agric., No. 7, 40 p.

Winn, H. E. 1958. Comparative reproductive behavior and ecology of fourteen species of darters (Pisces-Percidae). Ecol. Monogr. 28:155-191.

Woods, L. P., and R. F. Inger. 1957. The cave, spring, and swamp fishes of the family Amblyopsidae of the central and eastern United States. Amer. Midland Nat. 58:232-256.

Woolcott, W. S. 1953. Some percid fishes of certain Tennessee counties. J. Tenn. Acad. Sci. 28:245-246.

Woolman, A. J. 1892. Report of an examination of the rivers of Kentucky, with lists of the fishes obtained. Bull. U. S. Fish Comm. 10:249-288.

_____. 1892. A report upon the rivers of central Florida tributary to the Gulf of Mexico, with lists of the fishes inhabiting them. Bull. U. S. Fish. Comm. 10: 293-302.

Zorach, T., and E. C. Raney. 1967. Systematics of the percid fish, *Etheostoma maculatum* Kirtland, and related species of the subgenus *Nothonotus*. Amer. Midland Nat. 77:296-322.

AMPHIBIANS AND REPTILES

Adler, K. K. and D. M. Dennis. 1962. *Plethodon longicrus*, a new salamander (Amphibia: Plethodontidae from North Carolina. Ohio Herp. Soc. Spec. Publ. 4: 14 p.

Babbit, L. H. 1937. The amphibia of Connecticut. Conn. State Geol. and Nat. Hist. Surv. Bull. 57:1-50.

Babcock, H. L. 1919. Turtles of New England. Mem. Boston Soc. Nat. Hist. 8:325.

_____. 1929. The snakes of New England. Boston Soc. Nat. Hist. Guides. No. 1. 30 p.

Bailey, J. R. 1937. Notes on plethodont salamanders of the southeastern United States. Occas. Papers Mus. Zool. Univ. Michigan, No. 364:1-10.

Barbour, T. 1926. Reptiles and amphibians: Their habits and adaptations. Houghton Mifflin, Boston. 125 p.

Bellairs, A. d'A. 1960. Reptiles: Life history, evolution and structure. Harper and Brothers, N. Y. 192 pp.

Bishop, S. C. 1927. The amphibians and reptiles of Allegany State Park. N. Y. State Mus. Handbk. 3:1-141.

_____. 1941. The salamanders of New York. N. Y. State Mus. Bull. 324:1-365.

_____. 1943. Handbook of salamanders. Comstock, N. Y. 555 p.

Blair, A. P. 1957. Part 3. Amphibians. Pg. 211-271. In: Blair, *et al.*, Vertebrates of the United States. McGraw-Hill, N. Y. 819 p.

Blanchard, F. N. 1922. The amphibians and reptiles of Western Tennessee. Occ. Papers Mus. Zool. Univ. Mich., No. 117. 18 p.

_____. 1925. A collection of amphibians and reptiles from southern Indiana and adjacent Kentucky. Papers Mich. Acad. Sci. 5:367-398.

Blatchley, W. S. 1901. On a small collection of batrachians from Tennessee, with descriptions of two new species. 25th Ann. Report Dept. Geol. Indiana. 1900: 759-763.

236

Brandon, R. A. 1966. Systematics of the salamander genus *Gyrinophilus*. Ill. Biol. Monogr. 35:1-86.

Brimley, C. S. 1907. Artificial key to the species of snakes and lizards which are found in North Carolina. J. Elisha Mitchell Sci. Soc. 23:141-149.

_____. 1912. Notes on the salamanders of the North Carolina mountains with descriptions of two new forms. Proc. Biol. Soc. Wash. 25:135-140.

_____. 1926. Revised key and list of the amphibians and reptiles of North Carolina. J. Elisha Mitchell Sci. Soc. 42:75-93.

_____. 1939-1941. The amphibians and reptiles of North Carolina. Carolina Tips, Vol. 2, Nos. 1-7, Vol. 3, Nos. 1-7, Vol. 4, No. 1.

Brooks, M. 1945. Notes on amphibians from Bickle's Knob, West Virginia. Copeia, 1945:231.

Cagle, F. R. 1941. A key to the reptiles and amphibians of Illinois. Mus. Nat. and Soc. Sci. Contr. 5. 32 p.

_____. 1942. Herpetological fauna of Jackson and Union Counties, Illinois. Amer. Midland Nat. 28:164-200.

_____. 1957. Part 4. Reptiles. Pg. 273-358. In Blair, *et al.*, Vertebrates of the United States. McGraw Hill, N.Y. 819 p.

Cahn, A. R. 1937. The turtles of Illinois. Ill. Biol. Monogr. 16. 218 p.

Carr, A. F., Jr. 1940. A contribution to the herpetology of Florida. Univ. Fla. Publ. Biol. Sci. 3. 118 p.

Carr, A. 1952. Handbook of Turtles. Comstock, Ithaca. 542 p.

Carr, A., and C. J. Goin. 1955. Guide to the reptiles, amphibians, and fresh-water fishes of Florida. Univ. of Florida Press., Gainesville, Fla. 341 p.

Chermock, R. L. 1952. A key to the amphibians and reptiles of Alabama. Geol. Surv. Ala. Mus. Pap. 33. 88 p.

Conant, R. 1938. The reptiles of Ohio. Amer. Midland Nat. 30:1-200.

_____. 1952. Reptiles and amphibians of the northeastern states. Zool. Soc. of Phil. 40 p.

_____. 1958. A field guide to reptiles and amphibians of the eastern United States and Canada. Houghton-Mifflin, Boston. 366 p.

_____. 1963. Evidence for the specific status of the water snake *Natrix fasciata*. Amer. Mus. Nov. No. 2122, 38 p.

Conant, R., and W. Bridges. 1939. What snake is that? Appleton-Century-Crofts, N.Y. 163 p.

Cook, F. A. 1942. Alligators and lizards of Mississippi. Bull. Miss. State Game and Fish Comm. 20 p.

_____. 1954. Snakes of Mississippi. Bull. Miss. State Game and Fish Comm. 40 p.

_____. 1957. Salamanders of Mississippi. Bull. Miss. State Game and Fish Comm. 28 p.

Cope, E. D. 1889. The Batrachia of North America. U.S. Nat. Mus. Bull. 34:1-525.

_____. 1900. The crocodilians, lizards, and snakes of North America. U.S. Nat. Mus. Rept. for 1898. 1117 p.

Corrington, J. D. 1929. Herpetology of the Columbia, South Carolina, region. Copeia 172:85-88.

Curran, C. H., and C. Kauffeld. 1937. Snakes and their ways. Harper, N.Y. 285 p.

DeKay, J. E. 1842. Zoology of New York. Pt. 3. Amphibians and reptiles. White and Visscher, Albany. 104 p.

Dickerson, M. C. 1906. The Frog Book. Doubleday, Page and Co., N.Y. 253 p.

Ditmars, R. L. 1905. The reptiles of the vicinity of New York City. Amer. Mus. Jour. 5:93-140.

_____. 1931. Snakes of the world. Macmillan. N.Y. 207 p.

_____. 1936. The reptiles of North America. Doubleday, Garden City. 476 p.

_____. 1939. A field book of North American Snakes. Doubleday and Co., N.Y. 305 p.

Dunn, E. R. 1918. A preliminary list of the reptiles and amphibians of Virginia. Copeia. 53:16-27.

_____. 1926. The salamanders of the family Plethodontidae. Smith College Anniversary Ser. 7. 441 p.

Eckel, E. C., and F. C. Paulmeier. 1902. Catalogue of New York reptiles and batrachians. N. Y. State Mus. Bull. 51:355-414.

Fowler, H. W. 1907. The amphibians and reptiles of New Jersey. Ann. Rept. N. J. State Mus. 1906:23-348.

_____. 1925. Records of amphibians and reptiles for Delaware, Maryland and Virginia. I. Delaware. Copeia (145):57-61.

Gadow, H. F. 1909. Amphibia and reptiles. Cambridge Nat. Hist. Vol. 8. 558 p.

Garman, H. 1892. A synopsis of the reptiles and amphibians of Illinois. Bull. Ill. State Lab. Nat. Hist. Vol. 3. 175 p.

Goin, C. J., and O. B. Goin. 1962. Introduction to herpetology. W. H. Freeman and Co., San Francisco. 341 pp.

Grobman, A. B. 1944. The distribution of the salamanders of the genus *Plethodon* in Eastern United States and Canada. Ann. N. Y. Acad. Sci. 45:261-316.

_____. 1949. Some recent collections of *Plethodon* from Virginia with the description of a new form. Proc. Biol. Soc. Wash. 62:135-142.

Hairston, N. G. 1951. Interspecies competition and its probable influence upon the vertical distribution of Appalachian salamanders of the genus *Plethodon*. Ecology 32:266-274.

Haltom, W. L. 1931. Alabama reptiles. Ala. Must. Nat. Hist. Pap. No. 11. 145 p.

Hay, O. P. 1892. The batrachians and reptiles of the State of Indiana. Ann. Rept. Dept. Geol. Ind. No. 17. 204 p.

Hay, W. P. 1902. A list of the batrachians and reptiles of the District of Columbia and vicinity. Proc. Biol. Soc. Wash. 15:121-145.

Henshaw, S. 1904. Fauna of New England. 1. List of the Reptilia. Occ. Pap. Bost. Soc. Nat. Hist. Vol. 7. 15 p.

Hensley, M. 1959. Albinism in North American amphibians and reptiles. Mich. State Univ. Biol. Ser. Publ. 1:135-159.

Highton, R. 1961. A new genus of lungless salamander from the coastal plain of Alabama. Copeia 1961:65-68.

_____. 1962. Revision of North American salamanders of the genus *Plethodon*. Bull. Fla. State Mus. 6:235-267.

Hoffman, R. L., and H. I. Kleinpeter. 1948. Amphibians from Burke's Garden, Virginia. Amer. Midland Nat. 39:602-607.

Kelly, H. A., A. W. Davis and H. C. Robertson. 1936. Snakes of Maryland. Nat. Hist. Soc. Maryland (Spec. Publ.). 103 p.

King, W. 1939. A survey of the herpetology of Great Smoky Mountains National Park. Amer. Midland Nat. 21:531-582.

Lamson, G. H. 1935. The reptiles of Connecticut. Conn. Geol. Nat. Hist. Surv. Bull. 54. 35 p.

Livezey, R. L., and A. H. Wright. 1947. A Synoptic Key to the Salientian Eggs of the United States. Amer. Midland Nat. 37:178-222.

Loding, H. P. 1922. A preliminary catalogue of Alabama amphibians and reptiles. Geol. Sur. Ala. Mus. Pap. No. 5. 59 p.

Mansueti, Romeo. 1941. A descriptive catalogue of the amphibians and reptiles found in and around Baltimore City, Maryland. Nat. Hist. Soc. of Md. Proc., No. 7:56.

Martof, B. S. 1956. Amphibians and reptiles of Georgia. Univ. of Georgia Press, Athens. 94 p.

Martof, B. and R. L. Humphries. 1955. Observations on some amphibians from Georgia. Copeia 1955:245-248.

McCauley, R. H. 1945. The reptiles of Maryland and the District of Columbia. Privately printed, Hagerstown, Md. 194 p.

Minton, S. A., Jr. 1944. Introduction to the study of reptiles in Indiana. Amer. Midl. Nat. 32:438-477.

238

_____. 1954. Salamanders of the *Ambystoma jeffersonianum* complex in Indiana. Herpetologica 10:173-179.

Morse, M. 1904. Batrachians and reptiles of Ohio. Proc. Ohio Acad. Sci. 4:93-144.

Neill, W. T. 1954. Ranges and taxonomic allocations of amphibians and reptiles in the southeastern United States. Publ. Res. Div. Ross Allen's Reptile Inst. 1, No. 7: 75-96.

_____. 1957. Distributional notes on Georgia amphibians, and some corrections. Copeia 1957:43-47.

Netting, M. G. 1939. Hand list of the amphibians and reptiles of Pennsylvania. Bienn. Rept. Penn. Fish Comm. for 1936-1938. 26 p.

Noble, G. K. 1927. Distributional list of the reptiles and amphibians of the New York City region. Amer. Mus. Nat. Hist. Guide Leaflet 69. 9 p.

_____. 1931. The Biology of the Amphibia. McGraw-Hill, N. Y. 577 p.

Oliver, J. A. 1955. The natural history of North American amphibians and reptiles. D. Van Nostrand Co., Princeton. 359 p.

Oliver, J. A., and J. R. Bailey. 1939. Amphibians and reptiles of New Hampshire, exclusive of marine forms. Biol. Surv. Conn. Watershed Rept. 4. 195 p.

Parker, M. V. 1939. The amphibians and reptiles of Reelfoot Lake and vicinity, with a key for the separation of species and subspecies. J. Tenn. Acad. Sci. 14:72-101.

Perkins, C. B. 1942. A key to the snakes of the United States. Bull. Zool. Soc. San Diego 24. 79 p.

Pickens, A. L. 1927. Amphibians of upper South Carolina. Copeia 165:106-110.

Pope, C. H. 1937. Snakes alive and how they live. Viking Press, N. Y. 238 p.

_____. 1939. Turtles of the United States and Canada. Alfred A. Knopf, N. Y. 343 pp.

Pope, T. E. B., and W. E. Dickinson. 1928. The amphibians and reptiles of Wisconsin. Bull. Public Mus. Milwaukee 8. 138 p.

Reed, C. F. 1955. Notes on salamanders from western Connecticut, with especial reference to *Plethodon cinereus*. Copeia 1955:253-254.

Rhoads, S. N. 1895. Contributions to the zoology of Tennessee. No. 1, Reptiles and amphibians. Proc. Acad. Nat. Sci. Phil. 47:376-407.

Richmond, N. D. 1958. The status of the Florida snapping turtle. Copeia. 1958:41-43.

Romer, A. S. 1956. Osteology of the reptiles. Univ. of Chicago Press. 772 p.

Rose, F. L. 1963. A new species of *Eurycea* (Amphibia: Caudata) from the southeastern United States. Tul. Stud. Zool. 10:83-128.

Rossman, D. A. 1960. Herpetofaunal survey of the Pine Hills area of southern Illinois. Quart. Jour. Fla. Acad. Sci. 22:207-225.

_____. 1962. *Thamnophis proximus* (Say), a valid species of garter snake. Copeia 1962:741-748.

_____. 1963. Relationships and taxonomic status of the North American Natricine snake genera *Liodytes*, *Regina*, and *Clonophis*. Occ. Pap. Mus. Zool., La. State Univ. 29. 29 p.

Raugh, R. 1951. The Frog: Its reproduction and development. Blakiston, Phil. 336 p.

Ruthven, A. G., C. Thompson and H. T. Gaige. 1928. The herpetology of Michigan. Mich. Handbook Ser. Univ. Mich. 3. 228 p.

Schmidt, K. P. 1924. A list of amphibians and reptiles collected near Charleston, S. C. Copeia 1924:67-69.

_____. 1953. A check list of North American amphibians and reptiles. Amer. Soc. Ichthyologists and Herpetologists. 280 p.

Schmidt, K. P., and D. D. Davis. 1941. Field book of snakes. G. P. Putnam's Sons, N. Y. 365 p.

Schmidt, K. P., and W. L. Necker. 1935. Amphibians and reptiles of the Chicago region. Bull. Chic. Acad. Sci. 5:57-77.

Schwartz, A. 1957. Chorus Frogs (*Pseudacris nigrita* LeConte) in South Carolina. Amer. Mus. Nov. No. 1838:1-12.

Sinclair, R. M. 1950. Some noteworthy records of amphibians and reptiles in Tennessee. Herpetologica 6:200-202.

Smith, H. M. 1946. Handbook of lizards. Comstock, Ithaca. 557 p.

Smith, P. W. 1961. The amphibians and reptiles of Illinois. Ill. Nat. Hist. Surv. 28: 1-298.

Smith, P. W., and S. A. Minton, Jr. 1957. A distributional summary of the herpeto- fauna of Indiana and Illinois. Amer. Midland Nat. 58:341-351.

Smith, P. W., and H. M. Smith. 1962. The systematic and biogeographic status of two Illinois snakes. Occ. Pap. C. C. Adams Center for Ecol. Stud. No. 5. 10 p.

Smith, W. H. 1882. Report on the reptiles and amphibians of Ohio. Rep. Ohio Geol. Surv. 4:629-734.

Surface, H. A. 1906. The serpents of Pennsylvania. Zool. Bull. Div. Zool. Penn. State Dept. Agr. 4:113-208.

_____. 1907. The lizards of Pennsylvania. Zool. Bull. Div. Zool. Penn. State Dept. Agr. 5. 233 p.

Taylor, E. H. 1935. A taxonomic study of the cosmopolitan scincoid lizards of the genus *Eumeces* with an account of the distribution and relationships of its species. Bull. Univ. Kansas. 36:1-643.

Trapido, H. 1937. The snakes of New Jersey, a guide. Newark Mus. Spec. Publ. 60 p.

Uzzell, T. M., Jr. 1964. Relations of the diploid and triploid species of the *Ambystoma jeffersonianum* complex (Amphibia, Caudata). Copeia 1964:257-300.

Verrill, A. E. 1863. Catalogue of the reptiles and batrachians found in the vicinity of Norway, Oxford Co., Maine. Proc. Bost. Soc. Nat. Hist. 9:195-199.

Wake, D. B. 1966. Comparative osteology and evolution of the lungless salamanders, family Plethodontidae. Mem. S. Cal. Acad. Sci. 4:1-111.

Walker, C. F. 1946. The amphibians of Ohio. I. The frogs and toads. Ohio State Mus. Sci. Bull. Vol. 1:1-109.

Webb, Robert G. 1962. North American recent soft-shelled turtles. Mus. Nat. Hist., Univ. Kans. Publ. 13:429-611.

Weber, J. A. 1928. Herpetolotical observations in the Adirondack Mountains, New York. Copeia 169:106-112.

Williams, R. S. 1933. Notes on some Kentucky amphibians and reptiles. Bull. Baker- Hunt Mus. Covington. 22 p.

Wright, A. H. 1929. Synopsis and Description of North American tadpoles. Proc. U. S. Nat. Mus., Vol. 75, No. 11:1-70.

_____. 1947. Life-histories of the frogs of Okefinokee Swamp, Georgia. Macmillan, N. Y. 497 p.

Wright, A. H., and S. A. Bishop. 1915. A biological reconnaissance of the Okefinokee Swamp in Georgia. The reptiles. Proc. Acad. Nat. Sci. Phila. 1915. 85 p.

Wright, A. H., and A. A. Wright. 1949. Handbook of frogs and toads of the United States and Canada. Comstock, N. Y. 640 p.

_____. 1957. Handbook of snakes of the U. S. and Canada. Comstock, Ithaca. 1105 p.

MAMMALS

Allen, G. M. 1904. Check list of the mammals of New England. Fauna of New England. Occ. Pap. Bost. Soc. Nat. Hist. 7:1-35.

_____. 1939. Bats. Harvard Univ. Press, Cambridge. 368 p.

_____. 1942. Extinct and vanishing mammals of the western hemisphere with the marine species of all the oceans. Am. Comm. Intern. Wildl. Prot. Spec. Publ. 11. 620 p.

Allen, H. 1893. A monograph of the bats of North America. U. S. Nat. Mus. Bull. 43. 198 p.

Allen, R. H., Jr., G. M. Kyle, and C. Peacock. 1954. Game, furbearing and predatory animals of Alabama. Ala. Dept. Cons. Div. Fish and Game. 35 p.

Anderson, E. P. 1951. The mammals of Fulton County, Illinois. Chic. Acad. Sci. Bull. 9:153-188.

Anderson, S. and J. K. Jones, Jr., Ed. 1967. Recent mammals of the world. Ronald, N. Y. 453 p.

240

Anthony, H. E. 1928. Field Book of North American Mammals. G. P. Putnam's Sons, N. Y. 625 p.

Arthur, S. 1928. The fur animals of Louisiana. La. Dept. Cons. Bull. 18, New Orleans. 444 p.

Asdell, S. A. 1964. Patterns of mammalian reproduction. 2nd ed. Cornell Univ. Press, Ithaca. 670 p.

Audubon, J. J., and J. Bachman. 1851. The viviparous quadrupeds of North America. V. G. Audubon, N. Y. 334 p.

Bailey, J. W. 1946. The mammals of Virginia. Williams Printing Co., Richmond. 416 p.

Bailey, V. 1895. The pocket gophers of the United States. U. S. D. A. Bull. Div. Ornith. and Mamm. 5:1-47.

_____. 1896. List of mammals of the District of Columbia. Proc. Biol. Soc. Wash. 10:93-101.

_____. 1897. Revision of the American voles of the genus *Evotomys*. Proc. Biol. Soc. Wash. 11:113-138.

_____. 1900. Revision of American voles of the genus *Microtus*. North Am. Fauna 17. 88 p.

Bangs, O. 1896. The skunks of the genus *Mephitis* of eastern North America. Proc. Biol. Soc. Wash. 10:145-167.

_____. 1896. A review of the weasels of eastern North America. Proc. Biol. Soc. Wash. 10:1-24.

_____. 1896. A review of the squirrels of eastern North America. Proc. Biol. Soc. Wash. 10:145-167.

_____. 1899. The land mammals of peninsular Florida and the coast region of Georgia. Proc. Bost. Soc. Nat. Hist. 28:157-235.

Beddard, F. E. 1902. Mammalia. Macmillan and Co., London. 605 p.

Blair, W. F. 1957. Part 6. Mammals. p. 615-774. In: Blair, Blair, Brodkorb, Cagle and Moore. Vertebrates of the United States. McGraw-Hill, N. Y. 819 p.

Bouliere, F. 1955. Mammals of the world, their life and habits. Knopf, N. Y. 223 p.

_____. 1956. The natural history of mammals. 2nd ed. Knopf, N. Y. 364 p.

Brimley, C. S. 1905. A descriptive catalogue of the mammals of North Carolina, exclusive of the Cetacea. Jour. Elisha Mitchell Sci. Soc. 21:1-32.

Brooks, D. M. 1959. Fur animals of Indiana. Ind. Dept. Cons. P. R. Bull. No. 4. Indianapolis. 195 p.

Brooks, F. E. 1911. The mammals of West Virginia. Rept. West Va. Bd. Agr. 20: 9-30.

Burt, W. H. 1946. The mammals of Michigan. Univ. Mich. Press, Ann Arbor. 288 p.

_____. 1957. Mammals of the Great Lakes region. Univ. Mich. Press., Ann Arbor. 246 p.

Burt, W. H. and R. P. Grossenheider. 1952. A field guide to the mammals. Houghton Mifflin, Boston. 200 p.

Cahalane, V. H. 1947. Mammals of North America. Macmillan, N. Y. 682 p.

Churcher, C. S. 1959. The specific status of the new world red fox. J. Mamm. 40: 513-520.

Clement, R. C. 1952. An annotated check-list of the land mammals of Rhode Island. Audubon Society of R. I., Providence. 4 p.

Cockrum, E. L. 1955. Reproduction in North American bats. Trans. Kansas Acad. Sci. 58:487-511.

_____. 1962. Introduction to mammalogy. Ronald, N. Y. 455 p.

Colbert, E. H. 1955. Evolution of the vertebrates. John Wiley and Sons, N. Y. 479 p.

Connor, P. F. 1953. Notes on the mammals of a New Jersey Pine Barrens area. J. Mamm. 34:227-235.

_____. 1960. The small mammals of Otsego and Schoharie Counties, New York. N. Y. State Mus. Sci. Serv. Bull. 382. 84 p.

Cory, C. B. 1912. The mammals of Illinois and Wisconsin. Field Mus. Nat. Hist., Chic. Zool. Publ. 11:1-505.

Coues, E., and J. A. Allen. 1877. Monographs of North American rodentia. Govt. Printing Office, Washington. 1091 p.

DeKay, J. E. 1842. Zoology of New York. Part I. Mammalia. White and Visscher, Albany. 146 p.

deVos, Anton. 1964. Range changes of mammals in the Great Lakes region. Amer. Midland Nat. 71:210-231.

Doutt, J. K., C. A. Heppenstall and J. E. Guilday. 1967. Mammals of Pennsylvania. Pa. Game Comm., Harrisburg. 281 p.

Dutcher, B. H. 1903. Mammals of Mount Katahdin, Maine. Proc. Biol. Soc. Wash. 16:63-71.

Ellerman, J. R. 1940. The families and genera of living rodents. Publ. Brit. Mus. Nat. Hist., London. Vol. 1. 689 p.

Elliott, D. G. 1901. A list of mammals obtained by Thaddeus Surber in North and South Carolina, Georgia and Florida. Field Columbian Mus. Zool. Series 3:31-57.

Evermann, B. W. and A. W. Butler. 1894. Preliminary list of Indiana mammals. Proc. Ind. Acad. Sci. 2:161-164.

Evermann, B. W., and T. H. Clark. 1911. Notes on the mammals of the Lake Maxinkuckee region. Proc. Wash. Acad. Sci. 13:1-34.

Flower, W. H., and R. Lydekker. 1891. An introduction to the study of mammals living and extinct. A. and C. Black, London. 763 p.

Gifford, C. L., and R. Whitebread. 1951. Mammal survey of south central Pennsylvania. Pa. Game Comm., Harrisburg. 75 p.

Glass, B. P. 1951. A key to the skulls of North American mammals. Dept. Zool. Okla. State Univ., Stillwater. 53 p.

Goldman, E. A. 1910. Revision of the wood rats of the genus *Neotoma*. North Am. Fauna 31. 124 p.

_____. 1918. The rice rats of North America (genus *Oryzomys*). North Am. Fauna 43. 100 p.

_____. 1950. Raccoons of North and Middle America. North Am. Fauna 60. 153 p.

Golley, F. B. 1962. Mammals of Georgia. Univ. Ga. Press, Athens. 218 p.

Goodwin, G. G. 1932. New records and some observations on Connecticut mammals. J. Mamm. 13:36-40.

_____. 1935. The mammals of Connecticut. State Geol. and Nat. Hist. Surv. Bull. 53. 221 p.

Gregory, W. K. 1910. The orders of mammals. Bull. Am. Mus. Nat. Hist. 27:1-525.

Grimm, W. C., and H. A. Roberts. 1950. Mammal survey of southwestern Pennsylvania. Pa. Game Comm., Harrisburg. 99 p.

Grimm, W. C., and R. Whitebread. 1951. Mammal survey of northeastern Pennsylvania. Pa. Game Comm., Harrisburg. 82 p.

Gunderson, H. L., and J. R. Beer. 1953. The mammals of Minnesota. Univ. Minn. Press, Minneapolis. 190 p.

Hahn, W. L. 1907. Notes on the mammals of the Kankakee Valley. U.S. Natl. Mus. Proc. 32:455-464.

_____. 1909. The mammals of Indiana. 33rd Ann. Rept. Dept. Geol. and Nat. Res. Ind. 419-654.

Hall, E. R. 1951. American weasels. Univ. Kans. Publs. Mus. Nat. Hist. 4:1-466.

_____. 1951. A synopsis of North American Lagomorpha. Univ. Kans. Publs. Mus. Nat. Hist. 5:119-202.

Hall, E. R., and E. L. Cockrum. 1953. A synopsis of North American microtine rodents. Univ. Kans. Publ. Mus. Nat. Hist. 5:373-498.

Hall, E. R., and W. W. Dalquest. 1950. A synopsis of the American bats of the genus *Pipistrellus*. Univ. Kans. Publ. Mus. Nat. Hist. 1:591-602.

Hall, E. R., and K. R. Kelson. 1959. The mammals of North America. Ronald, N.Y. 2 Vols.

Hamilton, W. J., Jr. 1930. Notes on the mammals of Breathitt County, Kentucky. J. Mamm. 11:306-311.

_____. 1933. The weasels of New York. Their natural history and economic status. Amer. Midland Nat. 14:289-344.

_____. 1939. American mammals. McGraw-Hill, N. Y. 434 p.

_____. 1941. Notes on some mammals of Lee County, Florida. Amer. Midland Nat. 25:686-691.

_____. 1963 Reprint. Mammals of eastern United States. Hafner, N. Y. 438 p.

Harper, F. 1927. Mammals of Okefinokee Swamp region of Georgia. Proc. Bost. Soc. Nat. Hist. 38:191-396.

_____. 1929. Mammal notes from Randolph County, Georgia. J. Mamm. 10:84-85.

_____. 1929. Notes on the mammals of the Adirondacks. Handbook N. Y. State Mus. 8:51-118.

Hollister, N. 1911. A systematic synopsis of the muskrats. North Am. Fauna 32. 47 p.

_____. 1913. A synopsis of the American minks. Proc. U. S. Nat. Mus. 44:471-480.

Howell, A. B. 1921. A biological survey of Alabama. I. Physiography and life zones. II. The mammals. North Am. Fauna. 45. 88 p.

_____. 1927. Revision of the American lemming mice (genus *Synaptomys*). North Am. Fauna 50. 38 p.

Howell, A. H. 1914. Revision of the American harvest mice (genus *Reithrodontomys*). North Am. Fauna 36. 97 p.

_____. 1915. Revision of the American marmots. North Am. Fauna 37. 80 p.

_____. 1918. Revision of the American flying squirrels. North Am. Fauna 44. 64 p.

_____. 1929. Revision of the American chipmunks (genera *Tamias* and *Eutamias*). North Am. Fauna 52. 157 p.

_____. 1938. Revision of the North American ground squirrels with a classification of the North American Sciuridae. North Am. Fauna 56. 256 p.

Jackson, C. F. 1922. Notes on New Hampshire mammals. Jour. Mamm. 3:13-15.

Jackson, H. H. T. 1915. A review of the American moles. North Am. Fauna 38. 100 p.

_____. 1928. A taxonomic review of the American long-tailed Shrews (genera *Sorex* and *Microsorex*). North Am. Fauna 51. 238 p.

_____. 1961. Mammals of Wisconsin. Univ. Wisc. Press, Madison. 504 p.

Jameson, E. W., Jr. 1949. Natural history of the prairie vole. Univ. Kans. Publ. Mus. Nat. Hist. 1:125-151.

Kellogg, R. 1937. Annotated list of West Virginia mammals. Proc. U. S. Nat. Mus. 84:443-479.

_____. 1939. Annotated list of Tennessee mammals. Proc. U. S. Nat. Mus. 86:245-303.

Kennicott, R. 1857. The quadrupeds of Illinois. Report of the Commissioner of Patents for the year 1856. House of Representatives Ex. Doc. No. 65. 8:52-110.

Kirk, G. L. 1916. The mammals of Vermont. Joint Bull. No. 2. Vermont Botanical and Bird Clubs, p. 28-34.

Komarek, E. V., and R. Komarek. 1938. Mammals of the Great Smoky Mountains. Bull. Chic. Acad. Sci. 5:137-162.

Krutzsch, P. H. 1954. North American jumping mice (genus *Zapus*). Univ. Kans. Publs. Mus. Nat. Hist. 7:349-472.

Layne, J. N. 1958. Notes on mammals of southern Illinois. Amer. Midland Nat. 60:219-254.

Lewis, J. B. 1940. Mammals of Amelia County, Virginia. J. Mamm. 21:422-428.

Lindsay, D. M. 1960. Mammals of Ripley and Jefferson Counties, Indiana. J. Mamm. 41:253-262.

Lowery, G. H., Jr. 1936. A preliminary report on the distribution of the mamals of Louisiana. Proc. La. Acad. Sci. 3:11-39.

Luttringer, L. A., Jr. 1931. An introduction to the mammals of Pennsylvania. Bd. Game Comm., Commonwealth Pa. Bull. 15. 62 p.

Lyon, M. W. 1904. Classification of the hares and their allies. Smithsonian Misc. Coll. 45:321-447.

Lyon, M. W., Jr. 1923. Notes on the mammals of the dune region of Porter County, Indiana. Proc. Ind. Acad. Sci. for 1923:209-221.

_____. 1936. The mammals of Indiana. Amer. Midland Nat. 17:1-384.

_____. 1942. Additions to the "Mammals of Indiana". Amer. Midland Nat. 27:790-791.

Merriam, C. H. 1895. Monographic revision of the pocket gophers, family Geomyidae (exclusive of the species of *Thomomys*). North Am. Fauna 8. 258 p.

_____. 1895. Revision of the American shrews of the genera *Blarina* and *Notiosorex*. North Am. Fauna 10. 34 p.

_____. 1884. The mammals of the Adirondack region. L. S. Foster, N. Y. 316 p.

_____. 1911. Synopsis of the weasels of North America. North Am. Fauna 11. 44 p.

Metzger, B. 1955. Notes on mammals of Perry County, Ohio. J. Mamm. 36:101-105.

Miller, G. S., Jr. 1897. Revision of the North American bats of the family Vespertilionidae. North Am. Fauna 13. 144 p.

_____. 1899. Preliminary list of the mammals of New York. Bull. N. Y. State Mus. 6:271-390.

_____. 1907. The families and genera of bats. Bull. U. S. Nat. Mus. 57. 282 p.

Miller, G. S., and G. M. Allen. 1928. The American bats of the genera *Myotis* and *Pizonyx*. Bull. U. S. Nat. Mus. 144. 218 p.

Miller, G. S., and R. Kellogg. 1955. List of North American recent mammals. Bull. U. S. Nat. Mus. 205. 954 p.

Mumford, R. E., and J. B. Cope. 1964. Distribution and status of the Chiroptera of Indiana. Amer. Midland Nat. 72:473-489.

Mumford, R. E., and C. O. Handley, Jr. 1956. Notes on the mammals of Jackson County, Indiana. J. Mamm. 37:407-412.

Necker, W. L., and D. M. Hatfield. 1941. Mammals of Illinois. Chic. Acad. Sci. Bull. 6:17-60.

Nelson, E. W. 1909. The rabbits of North America. North Am. Fauna 29. 314 p.

Norton, A. H. 1930. Mammals of Portland, Maine, and vicinity. Proc. Portland Soc. Nat. Hist. 4:1-151.

Odum, E. P. 1944. Notes on small mammal populations at Mountain Lake, Virginia. J. Mamm. 25:404-405.

Osgood, F. L., Jr. 1938. The mammals of Vermont. J. Mamm. 19:435-441.

Osgood, W. H. 1909. Revision of the mice of the American genus *Peromyscus*. North Am. Fauna 28. 285 p.

Palmer, R. S. 1954. The Mammal Guide. Doubleday, Garden City. 384 p.

Palmer, T. S. 1904. Index generum mammalium: A list of the genera and families of mammals. North Am. Fauna 23. 987 p.

Parker, H. C. 1939. A preliminary list of the mammals of Worcester County, Massachusetts. Proc. Bost. Soc. Nat. Hist. 41:403-415.

Patton, C. P. 1939. Distribution notes on certain Virginia mammals. J. Mamm. 20:75-77.

Peterson, R. L. 1966. The mammals of eastern Canada. Oxford Univ. Press, Toronto. 465 p.

Poole, E. L. 1932. A survey of the mammals of Berks County, Pennsylvania. Reading Public Mus. and Art Gall. Bull. 13:5-74.

Preble, E. A. 1899. Revision of the jumping mice of the genus *Zapus*. North Am. Fauna 15. 42 p.

Pruitt, W. O., Jr. 1959. Microclimates and local distribution of small mammals on the George Reserve, Michigan. Misc. Publ. Mus. Zool. Univ. Mich. 109:1-27.

Quay, W. B. 1949. Notes on mammals of Thomas County, Georgia, with two state records. J. Mamm. 30:66-68.

Rhoads, S. N. 1896. Contributions to the biology of Tennessee, No. 3. Mammalia. Proc. Acad. Nat. Sci. Phil. 48:175-205.

_____. 1903. The mammals of Pennsylvania and New Jersey. Philadelphia, privately printed. 266 p.

Richmond, N. D., and H. R. Roslund. 1949. Mammal survey of northwestern Pennsylvania. Pa. Game Comm., Harrisburg. 67 p.

Roberts, H. A., and R. C. Early. 1952. Mammal survey of southeastern Pennsylvania. Pa. Game Comm. 70 p.

Roslund, H. R. 1951. Mammal survey of north central Pennsylvania. Pa. Game Comm., Harrisburg. 55 p.

Sanborn, C. C. 1925. Mammals of the Chicago area. Field Mus. Nat. Hist. Leaflet 8:129-151.

Sanderson, I. T. 1955. Living mammals of the world. Hanover House, Garden City. 303 p.

Schmidt, F. J. W. 1931. Mammals of western Clark County, Wisconsin. J. Mamm. 12:99-117.

Schwartz, A., and E. P. Odum. 1957. The woodrats of the eastern United States. J. Mamm. 38:197-206.

Scott, W. B. 1913. A history of land mammals in the Western Hemisphere. Macmillan, N. Y. 786 p.

Seton, E. T. 1909. Lives of game animals. 4 Vol., Doubleday, Garden City.

Sherman, F. 1937. Some mammals of western South Carolina. J. Mamm. 18:512-513.

Sherman, H. B. 1936. A list of the recent land mammals of Florida. Proc. Fla. Acad. Sci. 1:102-128.

Simpson, G. G. 1945. The principles of classification and a classification of mammals. Bull. Am. Mus. Nat. Hist. 85. 350 p.

Taylor, W. P. 1956. The deer of North America. The white-tailed, mule and black-tailed deer, genus *Odocoileus*, their history and management. Stackpole Co. and Wildlife Mgt. Inst. 668 p.

Townsend, M. R. 1935. Studies on some of the small mammals of central New York. Roosevelt Wildlife Ann., 4:1-120.

Walker, E. P. ed. 1964. Mammals of the world. 3 Vol. Johns Hopkins Press, Baltimore.

Welter, W. A., and D. E. Sollberger. 1939. Notes on the mammals of Rowan and adjacent counties in eastern Kentucky. J. Mamm. 20:77-81.

Whitaker, J. O., Jr. 1967. Habitat and reproduction of some of the small mammals of Vigo County, Indiana, with a list of mammals known to occur there. Occas. Papers Adams Ctr. Ecol. Studies No. 16. 24 p.

White, J. A. 1953. Taxonomy of the chipmunks, *Eutamias quadrimaculatus* and *Eutamias umbrinus*. Univ. Kans. Publ. Mus. Nat. Hist., 5:563-582.

Wood, F. E. 1910. A study of the mammals of Champaign County, Illinois. Ill. Lab. Nat. Hist. Bull. 8:501-613.

Young, J. Z. 1957. The life of mammals. Oxford Univ. Press, London. 820 p.

Young, S. P. 1958. The bobcat of North America. Wildlife Mgt. Inst., Washington. 193 p.

Young, S. P., and E. A. Goldman. 1944. The wolves of North America. Wildlife Mgt. Inst., Washington. 636 p.

_____. 1946. The puma, mysterious American cat. Wildlife Mgt. Inst., Washington. 358 p.

SELECTED REFERENCES

Bishop, S. C. 1962, 1947. Handbook of salamanders. Hafner. Cornell Univ. Press. 555 p.

Burt, W. H. 1954. The mammals of Michigan. Univ. of Michigan Press 288 p.

Carr, A. 1952. Handbook of turtles: The turtles of the U. S., Canada, and Baja Calif. Cornell Univ. Press 542 p.

Carr, A. and C. J. Goin. 1959, 1955. Guide to the reptiles, amphibians and freshwater fishes of Florida. Univ. of Florida Press. 341 p.

Conant, R. 1958. A field guide to reptiles and amphibians of the United States and Canada east of the 100th Meridian. Houghton Mifflin Pub. Co. 366 p.

Hall, E. R. and K. R. Kelson. 1959. The mammals of North America. Ronald Press. 1083 p.

Hamilton, W. J., Jr. 1943. The mammals of eastern United States. Cornell Univ. Press. 432 p.

Hubbs, C. L. and K. F. Lagler. 1964, 1949. Fishes of the Great Lakes Region. Cranbrook Institute of Science. Univ. of Michigan Bull. 26. 186 p.

Schmidt, K. P. and D. D. Davis. 1941. Field book of snakes of the United States and Canada. Putnam Publishing Co. 365 p.

Smith, H. M. 1946. Handbook of lizards. Cornell Univ. Press. 557 p.

Smith, P. W. 1961. The amphibians and reptiles of Illinois. Illinois Natural History Survey Bull 28: 1-298.

Trautman, M. B. 1957. The fishes of Ohio. Ohio State Univ. Press. 683 p.

Wright, A. H. and A. A. Wright. 1949. Handbook of frogs and toads. Cornell Univ. Press. 640 p.

INDEX

252